Many Women Strong

A Handbook for Women Firefighters

Prepared by:
Women in the Fire Service
P.O. Box 5446
Madison, Wisconsin 53705
608/233-4768
608/233-4879 fax

Researchers & writers:
Brenda Berkman
Teresa M. Floren
Linda F. Willing

With assistance from:

Debra H. Amesqua	Kim Delgaudio	Carol Pranka
Freda Bailey-Murray	Patricia Doler	Andrea Walter
Joette Borzik	Julia Luckey	Grace Yamane

Introduction

Women have a long history as firefighters, yet only recently have significant numbers of women begun to choose the fire service as a career. While some volunteer fire departments have included women for years, many others are beginning to do so only now.

A woman considering volunteer service or a career as a firefighter often has many questions. What does the work involve? How can I best prepare myself for it? What are the challenges and rewards of firefighting? How will the fire department and the other firefighters accept me?

This handbook was created to help women who would like to become career, volunteer, or seasonal firefighters, as well as those who have just started on the job and are seeking guidance. It offers insights and suggestions from women who have been there: female firefighters, officers, and chiefs from all across the country. It attempts to present firefighting as it really is--neither glamorized nor trivialized--and to share answers to questions women commonly have about working in this still nontraditional field.

As a resource to fire service leaders, a companion document to *Many Women Strong: A Handbook for Women Firefighters* provides information about many management issues involved in the gender integration of the fire service. Copies of *Many Faces, One Purpose* are available from the U.S. Fire Administration.

Many Women Strong: A Handbook for Women Firefighters was prepared under contract to FEMA by Women in the Fire Service, Inc. It was made possible only with the assistance of dozens of people from fire departments and other agencies throughout the country who provided information and shared their valuable insights. For this assistance, the researchers and writers of this handbook offer their sincere thanks.

Women Firefighters: A Status Report

Nearly a quarter-century after women first entered firefighting as a career, more than 4,500 women hold career-level fire suppression positions in nearly a thousand fire departments in the United States. Hundreds more work for the Federal government or State agencies in wildland fire suppression roles. Women are career firefighters in Canada, Great Britain, France, Germany, the Netherlands, Denmark, Australia, New Zealand, South Africa, Costa Rica, Panama, and Brazil. Many thousands of other women, in the U.S. and elsewhere, work in the fire service in nonsuppression roles: emergency medical services (EMS), fire prevention, inspection, arson investigation, communications, and public education.

More than 900 U.S. fire departments employ women firefighters; however, a few major departments have yet to hire their first woman. While nationwide only about 2 percent of firefighters are women, many career departments' percentages are three or four times that, and a few departments' ranks are 10 to 15 percent female.

Women's history as volunteer firefighters is much longer than in the career sector, reaching back well over 100 years. While reliable numbers are difficult to obtain, it is estimated that among the volunteer and paid-on-call fire and rescue forces in the U.S. perhaps 40,000 are women firefighters, and still more women serve as emergency medical technicians (EMT's) and paramedics.

Women are found in all ranks of the fire service, from recruit firefighter to chief of department. Women fire chiefs lead organizations ranging in size from small volunteer departments to those that protect cities the size of Madison, Wisconsin, county departments such as Cobb County, Georgia, and comparable agencies within the wildland fire service. The first generation of career women firefighters, that entered the fire service in the mid- to late-1970's, is coming of age, and the number of career fire service women at the chief officer level increases every year.

There is no such thing as a "typical" woman firefighter. Women firefighters come from all backgrounds, races, and ethnicities. They may be single, partnered, married, divorced, or widowed. They may be 6'2" and weigh 200 pounds, or 5'1" and weigh 110 pounds. They may have no children, or be mothers or grandmothers. They may be as young as 18 or as old as 70. They may have a high school education or Ph.D. What this diverse array of women firefighters has in common is their dedication to their work and their commitment to serving their communities through the fire service.

Table of Contents

Becoming a firefighter

The fire service has changed a great deal in the past 30 years.

If you were a firefighter in the 1960's, your job in many departments consisted mostly of taking care of the fire apparatus and the station and, when there was a fire, going to it and putting it out. Your protective gear--probably a canvas or rubber coat, thigh-length boots, and a heavy leather helmet with no eye protection--would now be considered primitive and unsafe. You probably had nothing to protect your lungs from the smoke and heat of a fire; coughing and choking on toxic fumes, and sometimes throwing up afterwards, were just part of being a good, tough firefighter. Overall concern for the health and fitness of firefighters was minimal. If anyone exercised on duty, it was usually out of boredom or a personal desire to be stronger, and workouts usually were limited to lifting weights someone had brought in from home.

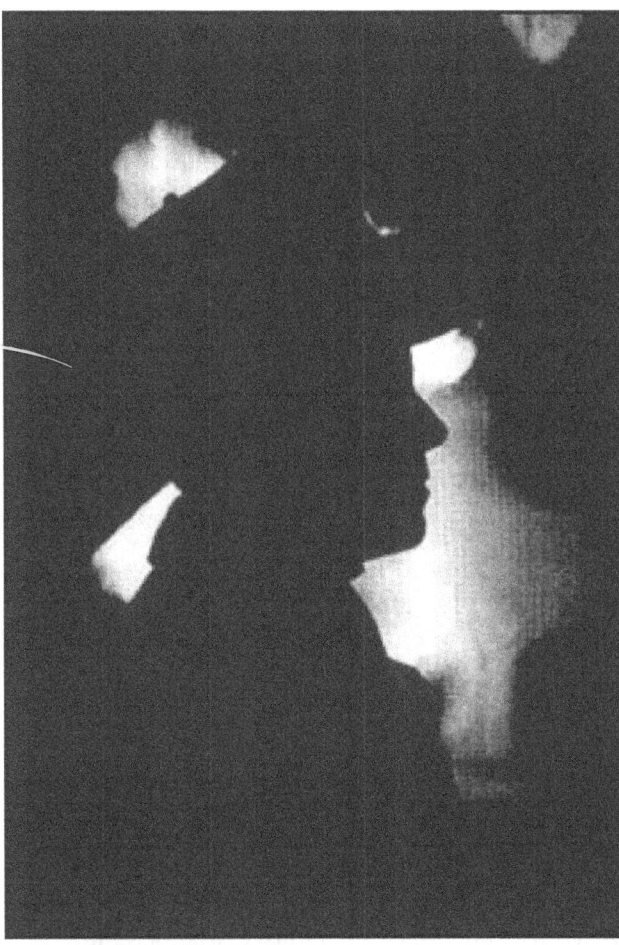

The job was more of an occupation than a profession. Other than fighting fires and, in some places, staffing load-and-go ambulances, you performed few community services. Most firefighters had a high school diploma at best; college and specialized fire service education were unheard of. Promotion to officers' and chiefs' positions came largely through seniority or through tests that measured your ability to memorize pages from designated books and pass a multiple-choice test based on that information.

By the late 1990's, almost all of this had changed. Firefighters in most fire departments now take part in public education, fire inspections, and other forms of community outreach. Almost all fire departments provide emergency medical response at the basic level, and many offer full-service paramedic care and patient transport. Special units of firefighters are trained to handle hazardous materials (haz mat) incidents, fast-water rescue, dive rescue, self-contained underwater breathing apparatus (SCUBA), and technical (high-angle and collapse) rescue. Arson investigation, fire code enforcement, and fire-safety education often form separate divisions within the fire department. A wide range of community-service careers has replaced the limited choices of only a generation ago.

Concern for firefighter safety has emerged in many ways. Protective breathing apparatus is a required part of the firefighter's gear on any incident where there might be a fire or toxic fumes. Accountability systems have been put in place to keep track of personnel at an incident. Most firefighters now wear devices that beep if the firefighter goes down or becomes trapped, to aid in a quick rescue. Firefighters have a much greater awareness of the fitness demands of the job, and many work out on duty as well as on their own time.

The field has become increasingly professional. It's not unusual for firefighters to have at least a 2-year degree in fire science or some other field, and chiefs of most major departments are expected to have master's degrees. Fire departments, colleges, and specialized training programs provide ongoing education in Command and management skills for Company Officers (CO's) and chiefs. Promotions in many fire departments are based on the employee's performance in a promotional assessment center instead of, or in addition to, more traditional types of tests and interviews.

The fire service workplace also has become more professional. Fire stations once boasted a fraternity-house atmosphere: they were "homes away from home" for men. Historically, this made sense: only four or five generations ago, firefighters worked six 24-hour days out of every seven and really **did** live at the fire station. Up until World War II, most departments used a 24-on/24-off schedule that persists in the Federal sector today. With so much of one's time spent in the station, and with work time encompassing aspects of domestic life such as cooking, eating, showering, and sleeping, it is not surprising that firehouses were viewed as very different from other workplaces. Drinking, sexual activity (with girlfriends or prostitutes), reading or viewing of pornography, and other traditional male social behaviors that would have been completely unacceptable in other work environments were commonplace in some fire stations.

Gradually, in most places, as firefighters' pay and their educational background have improved, such traditions have yielded to more enlightened management, and professional standards have replaced frat-house cultural norms in the fire service. Firefighters are now expected to behave like responsible public employees during their time on duty, and to treat the fire station like the workplace it is.

Because the fire service has changed, so has the way you should approach the possibility of becoming a firefighter. Back when firefighting was seen as manual labor, as semi-skilled work suitable to the sons of immigrant families, the idea of preparing for it as a career was unheard of. Today, anyone who applies for a firefighting job without preparing for it beforehand is unlikely to be hired. Much is expected of today's firefighters, and the competition for jobs is tough.

The impact of change

Organizational change happens slowly and unevenly. As might be expected, some fire departments have come farther than others; many have yet to make many of the changes mentioned above. On most departments, the most senior firefighters--those with more time on the job--were brought up in the old ways, and may or may not have adapted well to change. This resistance can cause conflict and resentment. Those who represent change, such as women firefighters, sometimes bear the brunt of this resentment. This isn't fair, and sometimes the resulting behavior is illegal, but women firefighters may find they have to deal with it nonetheless. *(See section on discrimination, page 27.)*

Why would I want to be a firefighter?

Firefighting isn't for everyone. People who are seriously afraid of heights or confined spaces, who don't function well in a crisis, or who in general would rather not introduce elements of danger into their lives, are unlikely to be attracted to the job. Some people prefer a typical business schedule to working 24-hour or other overnight shifts. (Most career firefighters work an average of 48 to 56 hours a week on 24-hour shifts.)* Other people may view the physical, manual-labor side of the job with distaste, preferring something more white-collar or intellectual.

---

*In the northeastern U.S., 10/14 or 9/15 shift schedules also are common. This means firefighters work a combination of 9- or 10-hour day shifts and 14- or 15-hour night shifts.

For those who do not see these aspects of firefighting as deterrents, it is an exciting, ever-changing, highly rewarding occupation. Most firefighters enjoy their work because of the warmth of camaraderie among the crew, the challenge of being able to bring physical skills and mental abilities to play in an emergency, and the opportunity to provide critical, life-saving services in a moment of need. Many also appreciate the work schedule, the job security in times of downsizing, and--in most fire departments--good pay and benefits.

What does it take to be a good firefighter?

If you are considering a career in firefighting, be aware that even though most firefighters are men, there are many ways to be a good firefighter, and you don't have to be male--or act like a man--to be good at the job. It is easy to feel out of place if all the firefighters you know or see are men. It may seem that, even if the door isn't officially closed to women, no woman could ever be enough like a male firefighter to be really good at the job. But what are the attributes of a good firefighter?

- honest and dependable;
- learns quickly, and can remember and use what he or she has learned when the pressure is on;
- physically fit: committed to a healthy lifestyle and to maintaining fitness;
- functions well as part of a team;
- cares about and respects co-workers and members of the community;
- communicates and listens well;
- dedicated to her/his work;
- has, and uses, common sense;
- emotionally stable and deals with stress appropriately;
- has a sense of humor; and
- open-minded and flexible, willing to try new things and listen to new ideas.

These attributes have nothing to do with gender, and no one person will possess them all. Each firefighter brings individual strengths to the team, and it is this variety of strengths that gives the team multiple options and balances out any individual weaknesses.

Women have functioned successfully as career firefighters and officers for more than 20 years, and as volunteers for much longer. Even if you're the first woman on your department, you're part of a strong tradition of women dedicated to the fire service.

Preparing for your career as a firefighter

Don't expect to become a firefighter tomorrow. Take a long-range approach to your firefighting career. Getting ready to become a firefighter should start years before you ever submit your first job application, ideally while you're still in high school. This will give you time to prepare yourself to be a good candidate for the job, as well as to decide if firefighting is right for you.

If you're still in school, get the basics in place: good reading comprehension and writing skills, computer literacy, basic math, and typing. Chemistry and biology, shop, auto mechanics, carpentry, and drafting will be useful. If your community (or the one where you hope to work as a firefighter) has a significant linguistic minority, study that community's language. In the U.S., this will usually be Spanish, but it also may be Hmong, Korean, Mandarin, or Vietnamese. It is useful for all firefighters to know the basics of American Sign Language (ASL), the language used by deaf and hearing-impaired people in every U.S. community.

Most women don't decide on a firefighting career until long after they've left high school. If you're already out of school, you can study most of the above subjects in a 2- or 4-year college program. Colleges in every State offer degrees in fire science or fire protection engineering, and some have residential firefighting programs that give students the chance to get hands-on experience and training. If you have a college degree, or if you know you'd eventually like to move up to a chief's position, a master's degree in public administration is an asset. Other courses that can be helpful for firefighters are public speaking, basic psychology, and government.

Education happens outside of school, too. The Red Cross in your community teaches first-aid and water safety classes; either it or the American Heart Association (AMA) probably offers cardiopulmonary resuscitation (CPR) training. Seek out opportunities to be trained as an EMT--volunteer fire departments and State agencies sometimes provide this training. If fire departments in your area give hiring priority to paramedics, and you're interested in becoming one, consider putting yourself through paramedic school. This means a significant investment of time and money, but it can enhance your chances of being hired considerably.

Depending on where you live, you may also have the option of putting yourself through basic firefighter training. In some States, you can only get this training once you've been hired by a fire department. Many departments put all newly-hired firefighters through recruit training, regardless of what prior training they may have. But in States such as Florida, where vocational schools and other agencies offer basic firefighter training to any interested student, it will be easier to get a firefighting job if you go through this training on your own.

Another way to get firefighter training, and to find out whether you like the work, is to become a fire cadet or a volunteer firefighter. Some fire departments have cadet programs or Explorer posts that allow high-school students to learn basic firefighting skills and spend time in the stations. For those over 18 (or sometimes 21), being a volunteer firefighter can provide excellent experience, education, and connections to job opportunities. *(See page* 13.)

If your area is served by a career-level fire department that does not have volunteer firefighters, find out if the department has any programs that involve other kinds of community volunteers, such as checking residential smoke detectors or teaching CPR. If you have an EMT or paramedic certification, check out options for volunteering in hospital emergency rooms.

Your physical training should be ongoing. Don't expect to sit in classes or behind a desk for years and then be able to get in shape after you've submitted your application. Actual firefighting takes up only a small percentage of the time that firefighters spend on duty, but that small percentage can demand extremes of strength and endurance from everyone involved. Despite many technological improvements in the equipment used, fighting a fire is still strenuous, hot, dirty, and often dangerous work.

Becoming a firefighter means a lifelong commitment to physical fitness: the earlier you make this commitment, the better. Get involved in sports teams, regular workouts, and other activities that will develop your strength and fitness and give you self-confidence in physically demanding situations. Your training routine should involve a weightlifting program as well as aerobic activities. (Be sure to get your doctor's okay before you begin any new exercise program.)

Most fire departments only give an entry-level (hiring) test every year or two, and it may be several months after that before the first recruit class is hired from the resulting list. While you're preparing yourself for your fire service career and waiting for your chosen fire department to announce a hiring opportunity, you'll probably have to work at another job. If you don't already have a job, look for other, nonfirefighting openings in the fire department or elsewhere in city or county government. These can provide you with excellent inside information, not only about job opportunities but about city government and the people involved. In addition, some towns and cities give hiring preference for other jobs to applicants who work for the municipality already.

Other kinds of jobs also can help you prepare to become a firefighter. Working as a dispatcher at the 911 alarm center is a good introduction to firefighters and their work. If you are a student, or for other reasons have summers available, consider applying for a seasonal job with wildland fire crews. Whatever your job, if it is not one that keeps you physically fit, be sure to integrate sports and other physical activities into your life outside of work.

Deciding what fire department you want to work for

Choosing where you want to work as a firefighter may be a decision you'll live with for the next 20 to 30 years, so it's worth making carefully. All fire departments are not the same, and they can differ in many ways. Obvious factors include the number of firefighters on the department, the size and location of the town or city, and the type of area protected.

You may already know what fire department you want to work for. Perhaps you already own a home and don't want to move, or you have friends on a particular department, or you simply have wanted to work for the XYZ Fire Department since you were little. Even if you think you're sure, it's worth reading this section to identify any questions you may want to ask.

Large department or small?

All sizes of fire departments have advantages and disadvantages. On a small fire department, you quickly come to know everyone on the job, as well as their families and friends. Departmental policies and the overall feel of station life often are less formal and militaristic than on larger departments. The department is more likely to be an integral part of the community, so the people whose houses you go to on fire or EMS calls are often friends or relatives of firefighters. Often you can follow up on the condition of patients you've treated, even after they leave the hospital, which is difficult or impossible in a big city. People you've dealt with on fires and other emergency calls may stop by the station later to visit or to thank you for helping them, bring you cookies during the holidays, and generally treat you as a person and not just as someone doing a job.

Fire departments in smaller towns usually are "slower," meaning they have fewer fires and other emergency calls. Whether this is a plus or a minus depends on what you want from your career. If you're itching to work at a station with lots of fire or EMS action, you probably will be disappointed on a small department. On the other hand, if you want a firefighting job that allows you an opportunity to be involved in many different kinds of activities, and to have your individual talents recognized and used to the fullest extent, you may be happier on a small department.

Promotional opportunities can be few and far between when a department has only a few officers. If your goal is to become a lieutenant or captain in the next few years, you may be better off on a department with more personnel. Keep in mind, though, that excellent career opportunities may be found on fire departments in towns or suburbs where the population is growing rapidly. The department will have to grow to keep up with the population, and those who are in it now will be prime candidates for promotion when it does.

Large fire departments, and particularly their busy stations, have status and glamour in the eyes of many firefighters who want as much action as possible. If an active station is important to you, you are more likely to find it in a larger department. Just keep in mind that every department has slow stations as well, and you won't necessarily be assigned to the busiest station just because you want to be.

Some advantages of larger departments include regular promotions and set promotional processes, more formalized training, and better training facilities. They also are more likely to have specialized teams such as haz mat, SCUBA teams, and technical rescue.

The disadvantage of a large fire department is that employees often get less individual recognition, just like employees of a big corporation. The status of working for a big-city department can be diminished if you end up feeling like just a small fish in a big pond. Bureaucracy and a many-layered hierarchy can make firefighters feel very remote from decisionmaking, which can be frustrating for some people. Depending on how station assignments are made, you can end up with a long commute across town to get to and from work. And, as a recruit, you often may find yourself "carrying your boots"--detailed to work at other stations--which can slow the process of integrating into your crew.

Where should the department be located?

If you want to settle down where you're now living, this question will be easy to answer. There may be only one fire department serving your area, or you may, at most, be able to choose from among a few. If you plan to move, you probably have some idea where you'd like to live: in the mountains, nearer to your family, in a warmer climate, by the ocean, etc. If you see yourself in a remote, rural setting, choosing a large metropolitan fire department may not make sense unless you don't mind a long commute in all kinds of weather, and the fire department doesn't have a residency requirement.

What do I want to do as a firefighter?

What fire department activities interest you most? If you hope to become a firefighter/paramedic, make sure the department you apply to has this as an option. Some fire departments will give you hiring preference or extra pay if you're already a paramedic when you're hired. In others, becoming a paramedic is a promotion you must compete for and earn only after you have some time on the job.

If you have a particular interest in specialty teams such as hazardous materials response, technical rescue, or SCUBA, be sure to choose a department that has those teams. Some fire departments participate in Urban Search and Rescue (US&R) teams that respond to major incidents all around the world. While positions on these teams are limited to a select, highly-trained few, you should choose a participating department if being on such a team is your goal.

If you want to be a firefighter only for a few years and then become a fire inspector or fire marshal, don't choose a department that has a civilian fire inspection bureau. Make sure the departments you apply to have these positions as promotions or in some other way part of a career ladder that begins in suppression.

The U.S. Forest Service and other Federal agencies, as well as several States, employ firefighters who specialize in wildland (forest fire) firefighting. (See pages 12-13.) Some urban fire departments, particularly in California, do both structural and wildland firefighting. If this is of particular interest to you, you should focus on these agencies and areas.

Answering the above questions should narrow down your choices. Once you have a short list of fire departments, you can start investigating the details of each one more closely. The questions that follow will help you gather the information you need to make your decision.

It may be that you don't really care **where** you work: you're unemployed or working at a job you hate, and you just want to get on a fire department and get started. In that case, you're probably already scanning the Sunday classifieds and putting in job applications every time you find an announcement for firefighter hiring. But even if you don't **think** you care, consider the questions below. If you have even two fire departments to choose from, you'll want to make the right choice.

Is this fire department a good place to work?

Go to the fire stations and talk with the firefighters and officers. (This will work best if you know someone on the department or can get a mutual friend to introduce you). How do they feel about their work? Do they seem to have respect for each other and for the department's management? Take a look at the station and the fire equipment. Is it clean, in good repair, and well taken care of? Or is it run-down and shabby through neglect or lack of funds?

What is the department's track record with respect to hiring women firefighters? The environment can be very different in a department where women have worked successfully for many years and earned the respect of their coworkers, compared to one that has yet to hire its first woman. Often the first woman in any fire department will encounter obstacles that are not found, or are much smaller, in fire departments that employ women already, where someone else has blazed the trail and smoothed out some of the rough spots. If you aren't sure you're cut out for the dual role of pioneer and firefighter, you may wish to concentrate on fire departments where women already are well established as firefighters and officers.

How do the department's personnel treat you? Do you feel comfortable talking with them as a prospective firefighter? Do they give you information about how to apply for the job, about the testing process? If possible, visit more than one station, including at least one where a woman is working. How do women on the job talk about the department and their coworkers?

What about pay and benefits?

Naturally, you will want to find out how much firefighters on the department make. Also find out how long their work week is. Firefighters on different departments in the same county or area may have pay scales and working hours that differ by 20 percent or more. The competition will be greater for the jobs with higher pay and shorter hours, but the benefits also are greater.

Don't just compare starting pay. How much do firefighters with one year on the job make? Two years? What is top firefighter pay, and how long does it take to get there? Are raises based strictly on longevity, or are they merit raises? If the department has a bargaining unit, when was the last contract negotiated, and is firefighter pay expected to go up with the next contract?

Must fire personnel meet an annual fitness or performance standard in order to keep their jobs or their suppression assignments? What does this consist of, and what happens to people who don't meet the standard? Do people on the job find the standard is relevant to the job's demands or useful for diagnosing problems, or do they feel the annual test is stressful or unnecessary?

Are pay bonuses or incentives available? Do paramedics or engine drivers receive additional pay, and how are personnel selected for these spots? If you plan to continue your college education while you're working as a firefighter, what kind of tuition reimbursement or educational incentive pay is offered?

What kind of promotional opportunities are available? How often are promotions made, and what kind of testing is given? How many years as a firefighter are required for promotion? Is a college degree or other advanced education required or an advantage?

Is the department unionized? Departments with firefighters' unions typically offer better pay and benefits than nonunion departments, and the union is, or should be, an advocate for you if you have any problems on the job. How is the relationship between labor and management? Has the union been supportive of its women members? Does it have a human relations committee or diversity task force?

Taking tests for practice

Taking a fire department's entry-level test just for practice is a good idea to help make you a better test-taker. Even though written and physical tests for fire departments differ considerably, people usually improve by taking them. Going through a few tests gets you used to the process and helps you feel less nervous about it.

On the other hand, you run one significant risk: you may do well enough on a "practice" test to be offered a job. If so, you may find it very hard to say no, even though it isn't really where you wanted to work. The financial security of having a firefighting job right now can be very appealing in comparison with the possibility of maybe getting on your chosen department a year from now. It may just seem like bad luck or foolishness to turn down a job offer. If you aren't sure you can be firm in your resolve, don't take a test for a fire department you **know** you don't want to work for.

The exception would be if you plan to work there only a short while and then move on to another department. While fire department managers don't appreciate this, it is done, and many smaller fire departments that can't afford to pay well have a high turnover of young firefighters going on to larger and better-paying departments. The recruit classes of some of those larger departments often include many experienced firefighters. This may be the option that best suits you, especially if you desperately need a job. It is usually easier to be hired by a fire department if you're already working for another one, as long as your work record is good.

The negative side to getting on "just any" fire department is that if it is a very badly run department, you may get turned off to the fire service or may not be able to stand being on the job long enough to get hired elsewhere. Women particularly run risks if the department is badly managed and fails to control or punish ugly behavior such as sexual harassment. Working in a hostile environment can harm your mental health or even put your physical safety at risk. Discriminatory behavior by coworkers and managers can result in you receiving poor training and bad evaluations that will harm your chances of getting hired elsewhere. Most fire departments are not like this, but departments of this type can still be found. Working for one of them is not the best idea, even in the short term.

The application and testing processes

Make sure you know how to find out when your chosen departments will be hiring. Get on a mailing list to receive their job announcement, if possible, or watch the classified ads or other places where openings are posted. When a fire department on your list announces a hiring opportunity, get a copy of all the available information and make sure you understand everything on it: dates, deadlines, and qualification requirements. If you're not sure about something, make a phone call or a visit to get clarification.

Different departments use different application processes. Some will mail their application forms; others distribute them only in person, and sometimes only at a specified time and place. Find out what you need to bring when picking up your application, such as a driver's license or proof of residency. The department may limit the number of applications given out; if this is the case, plan to be on hand very early to wait in line.

Some fire departments, particularly in California, use lotteries to reduce the number of applicants. Instead of bearing the expense of testing all firefighter applicants (which sometimes number in the thousands for only a handful of positions), they randomly pick a set number of applicants to continue through the process. This is an unfortunate practice and justifiably frustrating to the candidate who has prepared herself or himself for a firefighting career and expects to be able to compete for a job on the basis of qualifications, not chance. Nonetheless, it is a fact of life for the present, and unless it is challenged successfully in court or abandoned for other reasons, there is little or nothing the candidate can do about it except try not to be too disappointed.

You may be charged a fee (usually $10-25) when you submit your application. Sometimes applicants with severe financial constraints can ask for this fee to be waived.

Find out everything you can about the hiring process. How many steps does it involve, and what are they? Many variations are possible, but a typical process will look something like this:

1. The test is announced and applications are accepted. (In some places, applications are taken only on the day of testing.)

2. The written test is administered to all applicants.

3. The physical performance test is administered, either to all applicants or to those who passed the written test. In the past, fire departments often held the written and physical tests on the same day, but this is becoming less common. If you are traveling a long distance to take the tests, you may have to be prepared to make several trips.

4. The fire department or a contracted agency conducts a background check on applicants.

5. Applicants are given a psychological evaluation. (Only a small percentage of fire departments use this step.)

6. A hiring board or the fire chief conducts one or more interviews with top candidates.

Candidates who make it through these steps successfully are placed on a hiring or eligibility list. The order in which names appear on the list and the rules that govern the order in which candidates are hired vary from place to place. The list may be kept for 1 year, 2 years, or longer, depending on local policies and needs. When the department is ready to hire from the list, it will make a conditional offer of employment to the selected applicants, and send them through a medical evaluation, which sometimes includes a drug screening. Only after all of this has the applicant earned the chance to be hired as a fire recruit. (Smaller departments, or those that hire only certified firefighters, may hire applicants as probationary firefighters rather than fire recruits.)

The job application

If the application form asks you to submit a résumé, have yours done professionally, and tailor it to the job. Do not submit extra material that was not requested. **Never** lie on a job application. If you are asked for information that you feel may be harmful to your chances of being hired, write an explanatory note or ask to make an appointment with someone in charge to explain the circumstances.

The written test

Written tests for entry-level firefighter candidates are usually general-knowledge, civil-service type tests. A few fire departments may still use tests on firefighting subjects. These usually are based on a study guide distributed in advance. In Canada, so-called mechanical aptitude tests are still given; in the U.S., these have for the most part been abandoned as not relevant to the job.

Talk with people who have taken the department's test before. Even though the exact questions will be different, the types of questions, the format, and the test administration process may be similar. Take advantage of any study groups or preparatory test-taking assistance the fire department may offer, particularly if you are uncomfortable with written tests.* If the department does not provide such assistance, you may be able to find programs at local community colleges or university outreach programs that give you tips and practice in test-taking and job-interview skills.

As with every stage of the process, know the details. Find out where the test will be given, what you must bring with you (a photo ID, for example), and what time it will start. Plan to get there early. If you're driving a long way or to an unfamiliar part of town, give yourself plenty of extra time to get there. If you arrive even 5-minutes late, you will probably not be allowed to take the test or continue in the hiring process. Do not rely on other applicants for information about the testing process; deal directly with fire department or other testing officials regarding any questions or unusual situations.

The physical performance test

Different fire departments use different kinds of physical performance tests, and the exact components vary. The test you take may be a basic assessment of strength and fitness using measures such as sit-ups and a mile-and-a-half run, or it may be made up of events simulating tasks that are done at fires, such as hose drags and ladder raises. Find out all you can about the test in advance. If the department has a videotape of its test available, watch it several times. Attend test practice sessions if they're offered, preferably well in advance of your testing date. This will give you time to re-evaluate your workout routine to address any problem areas. A trainer at your gym or a coach at school should be able to help you tailor your workout to the specific events of the test.

Find out how the test will be scored. Many fire departments' tests are pass-fail; know what you need to do (usually a time you must meet) in order to pass. Some fire departments factor the applicant's time on the physical test into an overall score that determines their place on the hiring list. In such cases, the faster you get through the test, the better your chances are of being hired, so you should be prepared for an all-out effort.

On the day of the test, arrive early, well-nourished and well-rested. Take appropriate clothing: sweats, good athletic shoes, gloves if required. If the testing process will be a long one, make sure you have water and high-energy snacks on hand. If the day is hot, make sure to drink plenty of water before you start the test. Warm up and stretch just as you would before any strenuous workout.

Despite the fact that everyone is competing against each other for the same few job openings, a strong camaraderie often develops among the applicants taking a firefighter test together. Most women find this camaraderie includes them, even if there are few women taking the test. Such support and encouragement, even from total strangers, can help you perform well.

Make sure you understand all the instructions for each part of the test. Ask questions if necessary for clarification. If applicants wear firefighter protective clothing while taking the test, make sure yours fits. Don't be hesitant to call attention to yourself in this way. Having a glove fall off or a helmet slip down over your eyes during the test may mean the difference between passing and failing. The fire department or other testing agency must make sure the test conditions are as similar as possible for all candidates, and must make gear available to fit candidates of all sizes.

---

*Candidates with dyslexia may be able to get assistance or other reasonable accommodation to help them take the test.

The interview

An interview for a firefighter position should be approached in much the same way as one for any other professional job. The days are long past when it was appropriate to show up in casual clothes or with a casual attitude. Business attire and a professional outlook will demonstrate that you are serious about a career in the fire service and respectful of the interviewers.

Many books on job hunting offer excellent advice on job interviews, and firefighter candidates should take advantage of these. They cover common-sense items that are overlooked appallingly often by job applicants (i.e., Don't chew gum during the interview) as well as many other areas of preparation, appearance, and behavior. For a reality check on personal mannerisms that you may not otherwise have noticed, have a friend put you through a simulated interview, and videotape it. Watch the tape to see if you've presented yourself as someone **you** would want to hire. Do you make good eye contact with the interviewer? Do you sit comfortably and confidently in your chair, or do you slouch and wriggle? Is your speech punctuated with "umm's" and "you know's"? Work to correct any weaknesses that show up.

Don't lie awake nights trying to memorize, word for word, the perfect answer to every possible question you might be asked; you'll drive yourself crazy and lose a lot of valuable sleep. Instead, think of the general kinds of questions interviewers are likely to ask, and know the points you want to cover in your reply. Why do you want the job? What are your strengths? What are your weaknesses?

Be sure you've done your homework about the fire department. Know what services it provides, what community activities it's involved in, what programs it's proud of. Talking in your interview about how you'd like to become a firefighter/paramedic won't impress the interview panel if the fire department has just contracted its paramedic service out to a private company. Identify the ways your particular background and skills can make you an asset to this fire department, and bring these up in the interview.

After the tests are over

If you have applied for a job on a large fire department, be patient. It can take a long time to get all the applicants through the process, and when you're waiting to hear about a job, it will seem even longer. If you're dealing with a smaller department, particularly one where women are not yet welcome as firefighters, you may need to be a little more aware of how things are progressing. There's a fine line between staying on top of things and making a nuisance of yourself, but you should stay alert to make sure the process doesn't bypass you. If you have questions about how things are being run, and particularly if you learn that candidates lower on the hiring list have been offered jobs when you have not, ask. Get answers in writing, if possible.

Unless there is just one fire department you want to work for, keep putting your application in and taking tests with other departments while you're waiting to hear from the first one. You can always withdraw from a hiring process if you get a job elsewhere, and you may end up in the enviable position of having two job offers to choose from.

While most fire departments have professionalized their hiring practices and are careful to avoid unfair treatment of any applicant group, illegal discrimination still does crop up in hiring processes from time to time. If you believe you have been treated unfairly based on your race, gender, or religion, you may wish to pursue the matter through legal channels. (*See section on discrimination, page 27.*)

11

Getting a job as a Federal wildland firefighter

Following is some general information that will help get you started if you are looking for a Federal wildland firefighter position. Most of these jobs are with the U.S. Forest Service, the Bureau of Land Management (BLM), and the National Park Service. The U.S. Fish and Wildlife Service also has a few firefighter positions on national wildlife refuges.

Entry-level firefighters generally are hired as Forestry Technicians or Forestry Aides; a few positions are available for Professional Foresters specializing in fire management. These three titles are used in the Federal service to describe a myriad different positions for field-going individuals. For fire management positions, look at vacancy announcements that specify "Fire" in the job title, such as "Forestry Technician (Fire)" or "Forester (Fire)." Pay for these jobs varies by geographical area. Where the cost of living is high, employees receive specific locality pay.

Hiring procedures vary among agencies and even within each agency. If you are interested in working at a particular location, you should contact its personnel department and request information on their specific hiring procedures.

Most Federal wildland firefighters get permanent fire jobs only after a few seasons of temporary employment. Because competition is great, budget cutbacks can make it difficult to get hired permanently. Job opportunities sometimes vary due to politics, as it is the President and Congress who make the final decisions on the Federal fire budget. If you are persistent, however, you should get hired eventually.

How to find out about job openings

For 24-hour job information, call the U.S. Office of Personnel Management (OPM) at 912/757-3000. If you have a computer modem, you can dial 912/757-3100 for job information via OPM's electronic bulletin board. You can also reach the board through the Internet (Telnet only) at fjob.mail.opm.gov.

How to apply

Review the list of openings, decide which jobs you are interested in, and follow the instructions given. You may apply for most positions with a résumé, the Optional Application for Federal Employment, or any other written format. (You can get the Optional Application form by calling OPM or dialing its electronic bulletin board.) For jobs that are unique or filled through automated procedures, you will be given special forms to complete.

What to include on your application

Although the Federal government does not require a standard application form for most jobs, they do need certain information in order to evaluate your qualifications and determine whether you meet legal requirements for the job. If your résumé or application does not provide all the information requested in the job vacancy announcement, you may lose consideration for a job. Help speed the selection process by keeping your résumé or application brief and by sending only relevant material. Type or print clearly in dark ink. Be sure to include the following:

- Announcement number, title and grade(s) of the job for which you are applying.

- Personal information: name, address, phone numbers, Social Security number, country of citizenship, veteran's status, reinstatement eligibility if any, and highest Federal civilian grade held.

- Education: high school name, city, and State; date of diploma or GED; college/university name, city, and state; major; type of degree and year received. Include a copy of your college transcript if the vacancy announcement requests it.

- Work experience: job title, duties and accomplishments; employer's name and address; supervisor's name and phone number; starting and ending dates; hours per week and salary. Indicate whether they may contact your current supervisor.

- Other qualifications: job-related training courses, skills, certificates, and licenses; job-related honors, awards, and special accomplishments such as publications, memberships in professional or honor societies, leadership activities, public speaking, and performance awards.

Veteran's preference

If you served on active duty in the U.S. military and were separated under honorable conditions, you may be eligible for veteran's preference. To receive preference if your service began after 10/15/76, you must have a Campaign Badge, Expeditionary Medal, or a service-connected disability. For further details, call OPM at 912/757-3000. Select "Federal employment topics" and then "Veterans." You may access the electronic bulletin board at 912/757-3100.

Other tips

If you can't, or don't want to, move around in order to establish a work history with the Federal government, your chances of getting hired on permanently may be limited. Most Federal wildland firefighter positions are in the west. Oregon, Washington, and California in particular have lots of public land and thus need more wildland firefighters than States that have little public land. (Public land includes national forests, national parks or monuments, national wildlife refuges, and land managed by the BLM.)

If you're not the outdoors type, you are a poor candidate for jobs of this type. Wildland firefighting assignments take you to some of the most beautiful country in the U.S. However, this beautiful country is often remote and rugged, almost never a level hike, and dangerous to be in when it is burning.

If you've followed these suggestions and are still having problems getting hired, talk to someone who is already a Federal fire management employee about how he or she got hired. Also keep in mind that many States have their own wildland firefighter agencies, and employment with them might be another option.

Volunteer firefighting

Women who already have other jobs or careers often choose to get involved in the fire service by becoming volunteer or paid-on-call firefighters.* Most firefighters in the United States are volunteers, and rural areas and small towns in most parts of the country are protected by all-volunteer or predominantly volunteer fire departments. Many people who live in these areas find volunteering as a firefighter or EMT is a rewarding way to contribute to the community and provide a vitally needed service.

---

*Strictly speaking, "volunteer" firefighters receive no financial compensation for their work, while "paid-on-call" firefighters are, as the name indicates, paid for the time they spend on emergency calls and sometimes for training sessions. In practice, the term "volunteer" is used more loosely and generically, and does not always indicate whether the person receives money for her or his time.

Some larger towns and heavily populated suburban areas are served by "combination" fire departments: those with both career and volunteer personnel. These departments, particularly county-wide agencies in states such as Maryland and Virginia, can be quite large, with hundreds of personnel of each type.

Volunteer and paid-on-call firefighters normally perform the same emergency functions as career firefighters. The main difference is that volunteers usually respond to calls from home or work, although some departments require them to spend duty time at the fire stations.

For women who are considering a fire service career, becoming a volunteer can be an excellent first step. In exchange for your time, you will receive training and experience that will give you confidence and a good sense of whether firefighting is right for you. Whether you plan to move on to a career position or not, being a volunteer firefighter can be a challenging and enjoyable way to be involved in your community.

In choosing a volunteer fire department, you probably will have limited options unless you plan to move. Volunteer departments typically need members who either live or work in the area the department covers, although some may accept people from nearby areas. If you do have some choice in the matter, you should consider the following. Positive answers to most of these questions may indicate that you've found a good, progressive fire department where you can function well as a volunteer.

- How many hours a week or month are you required to commit to the department? Are you expected to respond on every call, or only on certain days or during certain hours?

- How do members of the department feel about its leadership? Is the department run well and efficiently? Are training and equipment as up to date as finances will permit? Do department personnel attend training sessions and conferences at State and regional levels?

- How is the work environment for women on the department? Do they receive the same levels of training and opportunities for promotion as the men? Do women feel welcome on the department and comfortable spending time at the station?

- What expenses does the department cover for its firefighters? Will they pay for your protective clothing (firefighting coat, pants, boots, helmet, and gloves), even if it has to be specially ordered to fit you, or do you have to buy it yourself? You may choose to join a department that can't or won't pay for these items, but you should know in advance what the costs to you will be.

- If this is a combination department, how is the relationship between career and volunteer personnel? Do volunteers receive adequate support and training, or are they treated like second-class citizens? What are the long-range plans for the department's staffing? If volunteers are being phased out, this may not be the best time to try to join.

The procedures and requirements for joining volunteer fire departments vary from place to place. In some, it is a simple application process; in others, existing department members vote to accept or reject new ones. A few volunteer departments have written or physical performance tests. It is wise to find out as much as you can about a department's selection process before entering into it.

Some men and women work as both career and volunteer firefighters, usually living in a small town or rural area (where they volunteer), and working their career job in a nearby city. Volunteer departments benefit from the expertise and experience of career personnel who wish to help out in this way. In the past, some combination departments used their personnel in a dual capacity, paying them as career firefighters for their normal duty days, and also allowing or requiring them to volunteer on their days off. The legality of this practice under the Fair Labor Standards Act is in dispute as of this writing.

Experience as a volunteer firefighter can be highly rewarding. Volunteer fire departments have a commitment and loyalty to their communities that are valuable, fruitful, and worthwhile. Many have been providing quality services for more than two centuries. Women have been an important part of many of these departments and will continue to be into the future.

Firefighting: what the job is like

The generation entering the fire service in the late 1990's is unique: it is the first generation in which women have grown up knowing they could become firefighters. Even so, most women who are now considering firefighting probably haven't thought of it as a career option for most of their lives, and may not really be familiar with what the job is all about.

Firefighting is a profession one does not enter into lightly. The decision to become a firefighter must be based on reality, and not on old myths, childhood images, or movie fantasies about what fighting fires or life in the fire station might be like. This section of the handbook will give potential firefighters some background to help them make the right decision about the job.

Women firefighters' experiences on the job vary considerably, depending on the particular fire department's history of employing women. The first woman hired by a department will be received with much more attention than will the fiftieth. A woman who is hired as a career firefighter by the town where she has been a volunteer for several years is likely to encounter little hostility; one who comes into a situation that, for right or wrong, has been polarized by lawsuits, usually will encounter a great deal of hostility. In part, how a woman is received also depends on her. The more informed she is about what to expect on the job, the better she will be able to meet its demands and adapt to new situations.

A day in the life of a firefighter

What is the job of firefighter really like? In describing the average day of a firefighter, first it's important to note that no two days can really be alike in the emergency services. Despite a framework that sets out rigid times for starting and stopping work, taking breaks and eating meals, the daily schedule is inherently subject to disruption, and those disruptions are the main reason firefighters are there. Contrary to popular belief, however, most of the day is not spent at fires and other emergencies: in fact, calls occupy less than 10 percent of the average firefighter's duty time.

The average work shift in the average fire station generally starts with an official or unofficial "roll call" at shift change to make sure everyone's there who's supposed to be, to communicate information from the off-going shift, to make assignments and to discuss the day's activities. Equipment check-in and placing one's fire gear on one's engine or truck usually follow immediately, often with a coffee break shortly afterwards. House cleaning is usually next. The rest of the morning may be taken up with inspections, training, or projects that can range from waxing the floor to building a new rack for the spare hose. Lunch is usually between 11:30 a.m. and noon and often is followed by a short break.

After lunch, inspections, training, or other projects resume. Training may involve vigorous, hands-on drills involving such things as hose lays, ladder work, and vehicle extrication, or (especially in cold or rainy weather) may be done in the classroom or by watching videotapes. Hands-on training usually will require a clean-up period afterwards, to reload hose or wash the vehicles. Other common daily activities include refueling, washing, and waxing vehicles; basic maintenance and checks on apparatus and equipment; inspections or walkthrough familiarizations of businesses and industrial properties in the district; territory drills (to make sure firefighters can find addresses and water supplies in their district); tours of the station for school classes or Scout troops, and so forth. Many departments have physical fitness equipment in the stations and allow or require firefighters to work out on duty.

After 4 or 5 p.m., one's time is generally one's own. Dinner is usually early, after which the kitchen is cleaned. For the rest of the evening, most firefighters watch television, play ping-pong, talk on the phone, read, study, or work on personal projects until going to bed. (Most departments prohibit firefighters from being in their bunks before a designated time usually 8 p.m. or later.) In some fire departments, friends or family may visit in the evening; in others, this is not allowed.

Fire departments approach weekends in varying ways. Most do not conduct weekend inspections, but some use the time for training, station cleaning, or other projects. Routine on holidays is typically very relaxed, with required activities limited to emergency responses.

Emergency responses

Most people join fire departments not for the daily routine, but for emergency activity: fire and/or EMS calls. What are those statistically small but often very exciting parts of the day like?

Just as most of the day is not spent on calls, most fire calls are not actual fires. Contrary to media images and other cultural myths, only one in 10 or 20 fire calls will involve flames. The actual percentage depends on the type of district the station covers. The majority of fire calls will be alarm and sprinkler system malfunctions, odors of smoke or gas with no actual fire, overheated fluorescent light ballasts or furnace blower motors, burned food on a stove, power lines arcing in trees, small gasoline spills, and the like. Fire engines (and in some places ladder trucks and other vehicles) also respond on EMS calls and car crashes, to assist paramedics and to provide safety and a faster response.

Of those fire calls that do involve actual fires, some will be small fires out on arrival; others will be leaf or brush fires, dumpster fires, or car fires. Still, it is the mystique of interior structural firefighting that lures most recruits to city fire departments, and it remains the psychological focus of the urban firefighter's job. Making an interior attack on a working fire (one that is actively burning and expanding) is an intense experience. Just seeing the glow in the night sky, or the smoke in the daytime, can quicken one's pulse. Active firefighting is physically demanding, frightening, and exhilarating. It is a challenge to overcome one's emotions and put all that one has learned into practice in an environment that is extremely hot, dark, usually noisy, and often confusing. The rewards come both through personal satisfaction and through having performed well as a crew. The part of one individual in the total scheme of things may be small, but the team has the satisfaction of working together to perform a difficult and important task. Bonding among firefighters is strong, and much of it is based on the intensity of the fireground experience.

Emergency medical calls bring their own rewards and frustrations. Fire dispatch centers receive more requests for EMS than for fire suppression, usually four or five times as many. This can keep firefighters from going stir-crazy in a slow station, or it can mean they must handle 10 or 15 calls a shift and end up physically fatigued and emotionally burned out. The frustrations of EMS include abuse of the ambulance service by people who use it as a taxi to the emergency room (which they're using in place of a doctor's office), the emotional strain of seeing people harm others or themselves (through violence, addiction, poverty, or lack of education), and the high percentage of trivial calls (the man who's had a stomach-ache for five days and calls you at 3 a.m.). The rewards are direct, personal, and vivid: successful interventions in life-threatening heart attacks and high-level trauma such as car wrecks, and the small but warm victories such as being able to calm a terrified child.

Firefighters respond to other kinds of emergencies as well. Incidents involving hazardous materials are on the increase, and firefighters may find themselves faced with spills or fires involving a wide range of toxic chemicals. Technical rescue teams, and sometimes engine and ladder companies, may be called on to rescue people trapped in cave-ins and collapses, or aboveground on scaffolding or other elevated locations. Some fire departments have water rescue teams: dive teams attempt to rescue people at risk of drowning in lakes and slow-moving water, while fast-water rescue teams handle similar situations in streams and rivers. And the entire department must be trained for ready response to the aftermath of rare but demanding disasters such as tornadoes, hurricanes, earthquakes, and--increasingly--terrorist activities such as bombings.

Though emergency response is the focus, 90 to 95 percent of one's life on the job is involved with other activities, mostly in the fire station. Women who leave firefighting jobs often do so because they find station life too hostile or uncomfortable, not because they couldn't fight fires or perform as a paramedic. Women considering firefighting careers should look beyond the glamour and challenge of emergency calls to how the rest of a firefighter's time is structured and spent.

Fire service mentality and traditions

Many women are unprepared for the paramilitary structure and mentality of the fire service. Most fire departments use a military rank structure, a strict chain of command, and a specifically delineated disciplinary process. (In some departments, for example, an employee who is late to work four times in one year is fired automatically.) The fire service is a male-constructed environment that women in most places have not so far had the time or numbers on the job to be able to change significantly. What this means for the recruit firefighter is learning to take orders without always receiving an explanation, having decisions made in an authoritarian rather than participatory way, and possibly even having to address officers as "Sir" or "Ma'am." For women coming directly from high school or college, the philosophy may be merely alien; for those coming from workplaces that are more democratic or humanistic, it can require some very difficult adjustments.

Part of the historical structure of the fire service is the tradition of treating new firefighters as less-than-equal participants in the workplace. This is more pronounced in some fire departments than in others. It can range from teasing and extra duties (raising and lowering the flag, answering the phone, refilling the soda machine) to harsh practical jokes, strict behavioral rules, and requiring recruits to spend their evening hours studying for probationary tests. Harsher forms of hazing generally are on the decline, especially in progressive fire departments, but some types are likely to persist for quite a while. Whether such behavior is right or wrong, it exists, and the individual recruit rarely is in a position to stop or avoid any except its most extreme forms. The probationary period is accepted largely as a time during which one is expected to perform the basic skills of the job competently without calling attention to oneself. Recruits are expected to be quiet, compliant, and receptive to criticism. While this attitude may be unfair or even counter-productive, one should expect to encounter it.

Generally, probationary firefighters are not covered by the union's contract with the employer (in fire departments that are unionized) and may not be permitted even to join the union. They thus have fewer safeguards regarding their employment and may be fired much more easily. All recruits should remember, however, that they never have to tolerate harassment based on their race, sex, or religion, even while they are on probation. The right to file a complaint over such problems, and not to be retaliated against for that reason, is protected by Federal and State law.

Recruit training

Fire recruit school can be more physically demanding than anything most women and many men have ever experienced. One woman firefighter noted that she had gone through Marine Corps boot camp, and that fire recruit school was tougher.[1] Training, in fact, often demands more from recruits on a sustained basis than anything the job will subsequently require. Women who are physically strong and aerobically fit will find their condition an advantage, if not a necessity, and still may be exhausted at the end of each day.

On the intellectual side, the challenges of recruit training are more variable, depending on the academy and on the individual's background and her academic abilities. It requires a great deal of memorization and pop-quiz recall of previously unfamiliar material, on subjects ranging from fire chemistry to hydraulics to the lengths of ladders carried on each of the department's trucks. As with probation in general, it is wise to show a low-keyed competence in recruit school: new recruits often are expected to know nothing, learn quickly, and speak only when spoken to.

Recruit school and the first weeks in the firehouse are stressful times for any new firefighter, and particularly for women who are among the first hired by the department. Female recruits should expect to depend heavily on the support of family and friends during this period, for practical concerns such as child care, meal preparation, and housework, and for the critical emotional support of a sympathetic listener and enthusiastic supporter.

Firefighting and women's home lives

Becoming a firefighter often has unexpected effects on a woman's social life. Many women firefighters find their career tends to dominate conversations or social gatherings away from work. Sometimes this is pleasant: it feels good to talk about something we care about or are proud of accomplishing. At other times, it may be tiresome, embarrassing, or inappropriate. The woman may feel as though she can't get away from work no matter where she goes, or that she is unusual in a negative way for having chosen such a career.

More importantly, the job very often is stressful to one's marriage or other primary relationship. The divorce rate among women firefighters is high, and the job almost always is a factor in the breakup. Many single women in the fire service have found their chosen profession alienates potential dates or partners, who often resent the amount of time and energy a firefighter's job demands. The stress of the job on a marriage or other relationship is one reason women end up dating or marrying other firefighters, who are much more likely to understand. Unfortunately, some fire departments do not look favorably on such relationships and will automatically place couples on separate shifts, which reduces by 50 percent or more the amount of time the couple can spend together. Fire departments often restrict promotional opportunities for married couples as well. (*See section on firefighter marriage, page 36.*)

The 24-hour work shift is a double-edged sword. Most firefighters prefer this shift and vigorously resist anything, even a promotion, that would move them to a 40-hour work week. Despite the fact that the resulting work week may be up to 56 hours long, the design of the shift permits more free time during weekdays than most jobs. Its disadvantages, however, include having to work many holidays and one day of most weekends, and having to arrange vacation or time trades in order to participate in classes or sports that meet on the same night each week. Being away from home for 24 hours at a time creates unusual child care needs for single parents, and it requires adjustments and flexibility on the part of one's partner and family. *(See section on child care issues, page 48.)*

Media attention to women firefighters

The first women to become firefighters on any department invariably draw media attention. A good chief or training officer will recognize the stress this creates for the women and will strictly limit the media's access to all recruits, preferably in a cooperative manner that keeps reporters from attempting to contact the women at home. Many women firefighters have found that media publicity, especially during the probationary period, hurts them on the job. They often find themselves misquoted and their stories sensationalized; despite this embarrassment, male coworkers often resent the attention given to the woman. Female recruits should familiarize themselves with the department's policy on speaking to the press, and should carefully consider whether they wish to take part in any interviews or photos not required by fire department administration.

Thousands of women in the U.S., Canada, and overseas currently work as firefighters, fire officers, and fire chiefs. Despite the obstacles they have had to overcome, the vast majority are in it for the long haul, enjoying productive, life-long careers in the field. They serve as role models and inspirations to the next generation of women hoping to enter the profession.

Notes:
[1] Unpublished survey data, Women in the Fire Service, 1985.

Seeing the whole job

In preparing to become firefighters, women often overlook the need to learn about the social environment and emotional demands of the fire station. The culture of a predominantly male workplace, the pressures that inevitably fall on the first women to work there, and the unique realities of fire and emergency response combine to create demands on the new female recruit that may catch her off guard. And while one cannot work out or go to school to prepare in this area, becoming aware of these demands can benefit women. Anticipating the internal and interpersonal dynamics that can exist or develop on the job allows a woman to strengthen her emotional "muscles" and create some multiple-choice "answers" for new and unfamiliar situations. Creating those choices and options, and helping to build strengths in new areas is what this section of the handbook is about.

Male culture in the fire station

The culture of the fire station is male in its language, values, skills, and traditions. This is true even though women have worked as career firefighters for more than 20 years. Some fire departments and stations are more progressive than others, and the most overt types of material and behavior (pornography, blatant sexual harassment) have been removed or reduced. Nonetheless, the vast majority of firefighters are men, and the culture that has grown up around firefighting reflects the traditional composition of its workforce.

Women joining fire departments often feel very much out of place at first: everything they know seems useless, and all the ways they behave seem wrong. If you look at the fire station as an environment dominated by male culture, you may find this alienation a little easier to handle: things are then not a matter of right and wrong, but of cultural differences. And just like any time you travel to another country, you need to know how to get along in a place that may be very different from your home. Becoming bicultural--able to communicate with and understand others from different backgrounds--is a positive challenge, as opposed to the desperate and unhappy feeling that you have to change who you are.

Firefighting and stereotypes

Firefighting usually is seen from the perspective of male values and "male" characteristics. Because fighting fires has traditionally been an all-male pursuit, when we try to assess what the job requires, we naturally focus on areas where men excel: where traditional male characteristics match skills and abilities needed on the job. Men value upper-body muscular development, aggressiveness, and emotional toughness in themselves, and since these are also valuable in firefighting, it's easy to think men must be better firefighters. But what about other traits that help one be an effective, healthy firefighter? Physical characteristics such as endurance, flexibility, and cardiovascular fitness, and skills such as the ability to manage emotionally-charged situations (injuries to a child, a fire involving deaths, or just the sadness and anger of a family watching its home burn) are also very important. It is a stereotype stemming from old firefighting traditions and an all-male workforce to believe that firefighting requires only those characteristics that are commonly thought of as masculine. Seeing the job of firefighter from a more neutral viewpoint can give women a better sense of where and how they fit in.

Women becoming firefighters should be aware not only of the bias inherent in traditional views of the job, but of the double bias that develops when these views combine with traditional ideas about women and men. One double stereotype, for example, says that firefighters must be tough, and women aren't tough. A common response to this is to believe it and either not become a firefighter (because you think you're not tough enough) or to become a firefighter anyway but feel you're inadequate at your job. Another response is to believe it but to make yourself an individual exception: "I'm different from other women; I'm tough and I can do it." But women do not have to buy into this double bind. The healthiest option is to realize that neither part of the stereotype is necessarily true. Gentleness and sensitivity are also important characteristics to bring to the job, and, whether tough or sensitive, many women are capable of being good firefighters.

Women firefighters often encounter the belief that the job requires abilities that men do not find desirable in women. Instead of saying that women can't be tough (or aggressive, or physically strong) this belief says that women who **are** those things aren't "real" women, in men's eyes. Sometimes this message comes from coworkers or others; sometimes it comes from within. Many women have grown up believing that "ladies" are, or should be, thin, delicate, and submissive. At some point, this ideal collides with the new image in the mirror of a muscular, strong and confident woman firefighter. This is painful for many women. Talking with women who are already on the job can help you learn to see yourself through your own eyes and those of people who care about you, not through old ideas about women.

It is perfectly possible that you as an individual will **not** fit your male coworkers' ideas of what women should be or can be. It's a shock to many women firefighters to find that they are working with men who have never had a female coworker (especially a peer) before. The change demands flexibility, adaptability, and a good sense of humor on everyone's part.

Interacting with men in the workplace

Women react in many ways to the firehouse workplace, depending on their individual personalities and backgrounds. Those who have worked in a male, blue-collar environment before, such as a factory or the military, already will have many of the skills that are useful in dealing with its pressures. Finding a friend to talk to who works in such a place or contacting women firefighters already on the job can give you the benefit of their experience.

In your first few days at your new fire station, things may feel very awkward. It takes time for any new recruit to be accepted, and sometimes it takes longer for women. Coworkers may range from friendly and helpful to resentful, aloof, or even unpleasant, depending on their personalities and on how they have been prepared for your arrival. Men who have not worked with women before may be afraid you will take joking, teasing, or any word of criticism as "discrimination." Let them know they can talk to you. Find points of interest for casual conversation, whether they're job-related or about something like food or sports. Be clear and consistent about your boundaries with respect to language, humor, or anything else. For example, if you don't care to discuss your private life but don't mind people making jokes about your small feet, let your coworkers know that.

Firehouse humor can be rough, abrasive, even crude and sarcastic, and it is often at someone else's expense. Some women find it very funny and are able to join in readily. Others find it painful, tiresome, and abusive. Humor is an important source of bonding among coworkers, but it has its limits. No one has to put up with jokes that are demeaning or insulting to them personally. To handle unwelcome humor, language, or conversation topics, it's best to be calm, not angry (if necessary, wait until later), clear and firm in explaining your feelings. Don't feel that because you've objected to one kind of humor, you can't continue to join in on joking that is not offensive.

Women often are alienated by the paramilitary aspects of the fire service. Firefighter training in particular tends to follow military and team-sports models. These styles are often effective with male recruits, most of whom have been through one or both of these experiences. They work less well in training women, who usually haven't. Women generally are not used to being yelled at as a routine form of correction, and they take it much more seriously and personally than male recruits do. Other forms of fire station or fireground direction, instruction, and education can feel equally harsh to women. Some fire departments are making an effort to modernize their teaching methods and make them more applicable to recruits from a variety of backgrounds. However, the tradition is well established and will take many years to change. Being prepared for it may make it less stressful for you.

Firefighters' work takes them into situations that can be highly emotionally charged, such as severe physical trauma and disfigurement, or the death of a child. Men often react to these situations in ways that are alien to women, and vice versa. After such an incident, men may seem to women to be callous, refusing to talk or perhaps indulging in gallows humor to relieve the emotional pressure. Women may instead **need** to talk about what happened and how they feel about it; men often will be uncomfortable with this and unwilling to engage. They also may resent the presence or perceived intrusion of female coworkers at a time when they feel vulnerable due to the emotional demands on them, and they may express feelings that clearly exclude the women. The passage of time will help you develop relationships with your coworkers that extend to-- and are strengthened by--such moments.

Pressures on the new woman firefighter

Almost every woman who becomes a firefighter feels the pressure to become "one of the guys." Many women, especially at first, consider it an honor, an undeniable mark of acceptance, for their coworkers to describe them in this way. Being "one of the guys" means adapting to male role-models, not standing out as different because you're a woman. The reward is an invisibility that can be a relief from the doubts and uneasiness that may have greeted you at first.

Starting out with a low profile, fitting in, and not making waves isn't a bad idea. No recruit firefighter, male or female, really wants to be visible, and most work hard for their peers' acceptance. But being "one of the guys" is a double-edged sword for a woman. It means acceptance, but it's an acceptance that, for many women, in the long run comes at too high a price. While some women truly enjoy being part of a male environment and fit into that role without having to change, many others find they must develop two personalities: one for work, and one for their real life. For the latter group of women, being accepted only as long as they act like men stops working when self-respect becomes more important than short-term survival.

This kind of acceptance may be comfortable and safe for a few months or years, but as your self-confidence on the job increases, you'll find you need to decide how much you will adapt to your surroundings and how much of yourself you can hang onto. Time will allow you to identify your priorities. You'll find it's okay to make waves when you have something important to fight for. Your ability to let go of being "one of the guys" will change over time; it does for everyone.

Some of the pressures of the job come not from coworkers, but from within yourself. It's common, for example, for a woman firefighter to feel she has to be perfect and can't afford to make a mistake. If her coworkers and officers doubt her abilities and expect her to have problems, this pressure is intensified. She may overcompensate by setting unrealistic standards for herself: she can never make a mistake, never ask for help. While this may seem admirable, someone who can't afford to make even one little mistake is, in reality, very insecure. Aspiring to excellence is fine, but a tense need for perfection is hard on everyone concerned.

If you find yourself falling into this pattern, stop and ask yourself why it isn't okay for you to do things wrong sometimes. No one is perfect. Recruits are expected to make mistakes, and you'll be no exception. What will happen when you do? Beware of internalizing other people's doubts about you. Take a direct approach to insecurities as they develop. Tackle directly the things that make you uneasy. If you think you have a weakness in a particular area, ask to work on that area until you feel confident with it.

There's a lot of truth to the saying that if you aren't making mistakes, you aren't learning anything. While the kind of mistakes that come from inattention, sloppiness, or carelessness only teach us to pay more attention or be more careful, other "mistakes" come from trying creative new solutions to problems or from being an eager learner. If you're hanging back and not trying new things until you're sure you can do them (or until no one else is watching), you're denying yourself a chance for valuable learning, the kind that comes from trial and error. Don't let your fear of being wrong keep you from creativity and experimentation, or from jumping in and trying a new skill.

Show your coworkers that you expect to be an equal member of the team, that you will do your share and expect them to do theirs. Often women will not ask for help with anything because asking seems like a sign of weakness. But firefighting is built on teamwork, and many tasks cannot be done (or done safely) by one person. It's hard not to want to "prove yourself" to a dubious coworker by carrying some unnecessarily heavy piece of equipment alone, but all this kind of behavior really proves is that there was some doubt in the first place. Develop the judgment it takes to know when you need to ask for help.

The other side of this is not to let yourself become less than a full member of the team. Your coworkers may seem very sure of themselves and their skills. Perhaps they jump aggressively into new tasks while you're still standing there trying to figure out what should be done. Or a well-intentioned coworker may try to take care of you by doing more than his share of the work--after all, he figures, it's easier for him. Although men may think they're doing you a favor by "helping" you out with hard or dirty tasks, you will learn nothing in this way. You may slide into a dangerous, "The guys will take care of it, so why should I bother?" attitude. From there, it is impossible to develop your skills and competence at the job, or to earn the respect of your coworkers.

If things are hostile at work, don't let the atmosphere of negative expectations undermine your self-confidence. It's easy to start thinking you know less than you do, or are capable of less than you are. Don't let pressures like these make you believe you're incompetent or keep you from participating fully in your job.

Boredom: a common enemy

Boredom is a real killer of motivation for both women and men in the fire service. Few firefighters are lucky enough to be assigned to a busy station as recruits, and many work most of their careers at stations that are downright slow. When you work at a slow station, it's easy to lose confidence in your skills. Training is all very well, but the new firefighter needs fires to remind her that, yes, she **can** put it all together for the real thing. Inactivity after a while also makes you lazy. The routine nature of the nonemergency work--checking the equipment, doing station maintenance--becomes boring and starts to seem unimportant, and after a while you don't want to do anything at all.

Tackle boredom aggressively. Find projects to do: busy-work is better than nothing, but even better are activities that will offer you and your coworkers a challenge. These can include anything from improving the way equipment is stored on the engine to learning sign language or developing a new public education program. For the evening hours, if you find yourself too often sprawled on the sofa watching reruns of boring movies, try to motivate yourself to bring in some projects from home, or offer to help a coworker with his or her project. Even if you know nothing at all about the work, most projects can make use of an assistant, and you may learn something in the process as well as creating a bond with your coworker. The point is to stay busy and stay challenged.

Take pride in your work and joy in your new skills. Don't let other people's negative expectations depress you into just sliding by, even if no one seems to recognize your competence. Don't let the boredom of a slow station draw you into laziness. Being a good, successful firefighter requires all the determination, motivation, and commitment you can bring to the job.

Interactions with other women at work

The majority of your coworkers, now and probably throughout your career, will be men. But it's becoming more likely, especially in larger fire departments, that you will have other women firefighters or officers in your station. It's virtually certain there will be other women on your shift in different stations or battalions. While this can be a great relief, it can also generate some unfamiliar pressures for both of you.

Women firefighters vary as much as the men on the job do. Sometimes it's hard not to be critical--or to feel criticized or undermined--when another woman on the department behaves differently from you. Realize and respect that other women must make their own decisions about how they behave in particular situations, how they interact with the men they work with, and how they feel about the job in general.

Male coworkers may try to draw you into criticizing other women, either in general or over a specific

situation. This can be a part of being "one of the guys," testing whether you're willing to side with them against her. Be very careful about situations like this. It's best not to express an opinion unless and until you have all the facts, including talking to the woman involved. When it's a serious situation, such as one involving sexual harassment, your support may be very important to her, and your opposition can be harmful. You and she may have very different ideas about what's harassment and what's acceptable behavior,

but you don't have to agree with how she feels about the incident in order to support her right to do something about it.

Generational differences often exist among women firefighters. Remember that the women who were the first to come onto the job have almost always had a tougher time of it than those who follow. This can help you understand if women with more seniority seem cool towards you. It's also useful for you to remember if you end up in that position 5 or 10 years from now and see new women recruits having an easier time of it. ("Well, when **I** was a rookie, **I** didn't get gloves that fit until after recruit school was finished!") Try to see things from the other woman's perspective, and offer her the respect she deserves for paving the way. Leave room for friendships to develop. Despite your differences, you have a great deal in common. Finding that common ground, and finding a source of support in each other, can be highly valuable.

Dealing with job discrimination and sexual harassment

Federal law (Title VII of the Civil Rights Act of 1964) and many State laws protect employees from being discriminated against based on their sex, race, or religion. If you feel you have been treated unfairly, and harmed by that treatment in some way, because you are a woman, African-American, Jewish, etc., you have the right to file a complaint with the Equal Employment Opportunities Commission (EEOC) aimed at reversing the harm that was done. Some States and many cities also have laws or ordinances prohibiting discrimination based on sexual orientation. If you believe you have been denied a job or otherwise treated unfairly because you are a lesbian (or because the employer thought you might be a lesbian), you may file a complaint with the appropriate State or city agency.

Deciding to file a complaint is never easy. The process is time-consuming and often unpleasant. Many women who have been through it have felt they were made into the "bad guys" even though they were the ones treated unfairly. Nonetheless, going through the complaint process and, if necessary, a lawsuit is sometimes the only way to get an employer to stop illegal behavior. Women who have filed complaints successfully against their fire departments often say they did it in part so other women wouldn't have to go through what they went through.

Photo copyright © Richmond Times-Dispatch. Used by permission.

The most common form of sex discrimination in the fire service is sexual harassment, which will be discussed at length below. This section will also deal with discriminatory hiring practices and other job problems that may be remedied through the EEOC complaint process.

Illegal treatment in the hiring process

When Federal laws against sex discrimination were first applied to the public sector in 1972, many fire departments were caught off guard. Some continued to evaluate job applicants in antiquated, haphazard ways, and for many years few regarded women applicants very seriously. Women who applied for firefighter positions in the mid-1970's often have memories of being laughed at and having their application forms refused or thrown in the wastebasket.

Most fire departments since then have made sincere efforts to professionalize their hiring processes and eliminate any illegal practices that have a negative impact on women or people of color. It is unlikely that anyone applying to a major metropolitan fire department in the 1990's will be confronted with blatant discriminatory treatment. Nonetheless, it is a possibility. Whether the department you're applying to is large or small, progressive or stuck in the Middle Ages, you should know what kinds of questions and treatment are against the law, and what your options are for dealing with such treatment if it should occur.

Being denied an application

It is extremely unlikely that any career-level fire department will refuse to give you an application on the basis of your gender. Fire departments that limit their applicant pool normally do so either on a first-come, first-serve basis, or through a lottery of all interested candidates' names. It is not legal for a career fire department or any other employer to say, "We won't let you apply for this job because you're a woman." (It may, however, in some States and under some conditions, be legal for volunteer fire departments to make a similar statement, as laws prohibiting employment discrimination do not always cover volunteers.)

Preference points

Some fire departments give hiring preference to military veterans, to residents of that particular city, or to people who currently work for the city. Even though veterans' points help many more men than women (because more men than women enter the military), they are not considered sex discrimination, because they are given to both male and female veterans. Given the Federal government's position against openly gay men and lesbians in the military, some have questioned whether veterans' preference might constitute illegal discrimination based on sexual orientation, in States and cities where such discrimination is against the law. This question has yet to be addressed by the courts.

A few part-paid, part-volunteer fire departments ("combination" departments) give preference to their own volunteers as candidates for career firefighter positions. Where this is the case, their selection process for volunteer firefighters should be looked at carefully; it probably will not be legal for such departments to deny women as a group the opportunity to become volunteer firefighters.

Irregularities in the testing process

The physical performance test is one area where illegal discriminatory treatment sometimes still occurs. It can be part of the test design, or it can occur in the process of administering and scoring the test.

Federal law requires all hiring tests to be job-related: that is, the agency giving the test must be able to show that it is linked to the actual demands of the job. The fire department or other agency must be able to demonstrate that people who do better on the test do better on the job, and people who do poorly on the test do poorly on the job. This job-relatedness will be difficult or impossible for the average firefighter candidate to determine as she or he is taking the test. When a test is thrown out for not being job-related, it is a court that does so, usually after one or more women applicants fail the test and bring a lawsuit.

What you will be able to observe when you take the test, and perhaps beforehand, is whether it is administered fairly and even-handedly. All applicants should, to the greatest extent possible, take the test under similar conditions and get an equal chance to do well on it. This means if the department provides firefighting gear such as helmets or gloves and requires the candidates to wear it during the test, the gear must fit everyone equally. It is a major disadvantage to have to take a physical test in gloves that are too big or a helmet that keeps falling off, and disadvantages linked to a person's smaller size usually are found to constitute sex discrimination.

It also means candidates must be given equal levels of encouragement or support before and during the test. If the people administering the test encourage male candidates as they go through the test, or allow other people to be nearby and provide encouragement, but women taking the test are treated with silence or with negative comments and hostile humor, there is a very clear difference in testing conditions between men and women. Most people whose job it is to administer entry-level tests are aware of this problem and will not let it happen. When the physical test is given under less formal conditions, however, such as at a fire station, using the station's regular duty personnel as monitors, discrepancies in treatment do sometimes arise.

Illegal questions in the job interview

Most questions pertaining to the applicant's gender, race, religion, and marital status are illegal. Under the Americans with Disabilities Act, some questions about physical handicaps also are illegal. In many cities and some States, prospective employers may not ask about an applicant's sexual orientation. The trend in general appears to be to limit the range of "intrusive" questions that may be asked on a job interview or pre-employment polygraph.*

Before you go into the interview, know what types of questions are illegal. For example, these questions should not be asked:

- "Are you married?"

- "What does your husband think about you becoming a firefighter?"

- "Do you have children, or plan to have children?"

- "What child care arrangements will you make if you get the job?"

- "Have you ever been divorced?"

- "What is your ethnic background?"

- "Are you a Jehovah's Witness?"

- "Are you HIV-positive?"

Questions such as the following, however, are legal:

- "Can you work every day of the week? "

- "Can you work 24-hour shifts?"

- "Do you have any physical limitations that would prevent you from doing the job of a firefighter?"

It is usually best to answer all interview questions, even if you think they're illegal. In the first place, you might be mistaken; if you refuse to answer a question that turns out to be legal, you will probably not get the job. If the question was illegal and you answer it and get the job anyway, it won't have mattered; once you're securely employed, you may want to take up the issue again in order to ensure that future applicants aren't asked similar questions. If you don't get the job, you may have grounds to file a charge against the employer.

---

*A 1995 U.S. District Court decision enjoined a municipality from asking police and fire applicants questions about consensual sexual activity "except to the extent the act was unlawful... and involved a minor and occurred within three years of the screening," extramarital sex, the use of marijuana more than six months previously, and numerous other areas.[1]

Under what conditions should you refuse to answer an illegal question on a job interview? If the question is completely offensive to you, or if you feel an honest answer to an illegal question would provide information that could be damaging to you after you're hired, then you may choose not to answer. This can be done tactfully and doesn't have to be an "in your face" assertion that you know your rights. Don't lie, even in response to outrageously illegal questions: if you answer a question, answer it honestly.

Many employers ask applicants to fill out a form that helps the employer track the numbers of women, people of color, and members of other definable groups who are applying for a particular job. This form will ask for your racial or ethnic background, gender, and age, among other things, but will not ask for your name. As long as the form is **voluntary, anonymous,** and **not attached to the application form**, it is usually legal for employers to use it.

Some fire departments use a psychological screening as part of the hiring process. Questions that would be illegal in an interview (especially regarding sexual activities or marital and family relations) should not be asked in the psychological evaluation unless the content of that discussion is guaranteed to be confidential. The psychological screening may not be used as a method of obtaining illegal information on an applicant if that information is simply going to be passed back to the employer.

What to do if it happens to you

Go into the hiring process with a positive but intelligent outlook, expecting no unfair or illegal treatment but prepared with the knowledge of what to do if it happens. Keep a journal of the entire process, with dates that you contacted people, the name of the person with whom you spoke, the dates you submitted your application and took each test, the scores you received, etc. Particularly document any aspects of the process that you felt were questionable or illegal. Who witnessed the behavior? Who else was affected by it? Your journal will probably turn out to have been just an exercise, and you will be able to throw it away later. But it's cheap insurance that will provide crucial documentation for your case if need be. Keep your notes accurate and to the point: if you end up in a lawsuit, they may not be considered confidential.

If you are rejected for employment, try to find out why. Was it a bad interview, too low a score on the abilities test, an unfavorable psychological evaluation? Get the explanation in writing if possible. If you are satisfied with it and with the way the hiring process went, use the information to improve your preparation process and work on your weak areas for next time.

If the explanation seems unreasonable, you noted irregularities in the hiring process, or you can't get them to give you an explanation, you may wish to pursue matters further. Get all the information you can. Talk with other women who went through the process and weren't hired. What reasons were they given? Did they feel the testing process was fair? What percentage of women passed the tests, compared to the percentages of men who passed? If you have friends who are currently on the department, what have they heard about the testing and hiring processes?

At this point, if it appears the process was conducted illegally, you may wish to file a charge with your State fair employment practices agency or the Federal EEOC, and/or consult an attorney. There are specific time limits (normally 6 months or 1 year) for filing a charge, so don't let things go on for too long while you decide what to do. Filing a charge doesn't cost anything, and you do not need an attorney in order to do so. If you can afford legal counsel, however, it's a good idea to get that guidance. Try to choose a lawyer who is experienced in employment discrimination law and who has handled previous cases involving public-sector employers. If other women applicants have been affected in the same way, you may be able to group together and pursue the matter in what is known as a "class action." This can be a stronger way of presenting your case, as well as providing support for the women involved.

Other forms of job discrimination

Discrimination can take many forms besides sexual harassment and being denied a job. It may become evident as a failure to promote a woman because someone in the department doesn't want women to be officers, or as a refusal to train women properly or consider them for special assignments. It may be as obvious as trying to fire a woman who becomes pregnant, or as subtle as giving women evaluations that are just never quite as good as the men's.

Whatever the form it takes, if you are being treated differently based on your race, sex, or religion, you have the right to follow the steps above and get the behavior stopped. You also have the legal right not to be retaliated against for filing a complaint.

Sexual harassment

Sexual harassment is a form of sex discrimination. It is illegal, it can devastate those who experience it, and it often destroys the morale and productivity of the work environment. It is common in the fire service, and the numbers are not getting better. As many as 85 percent of women firefighters have experienced some form of sexual harassment at work or in their volunteer departments.[2]

Sexual harassment is a power play that is degrading, humiliating, and intimidating to its victim. It is based on aggression and hostility, not sexual desire. The physical appearance and behavior of the victim do not cause harassment. Sexual harassment is not "natural attraction," a "compliment" to the victim, or a "normal" way for men to react to women in the workplace. It is a way men assert their dominance over women and, in some cases, try to force women to leave the job.* Compliments that are welcomed, attention that is mutually desired, and truly friendly jokes and teasing do not leave either participant feeling uneasy or intimidated. Sexual harassment does.

Sexual harassment is an intimate violation that often occurs without witnesses. Its victims generally feel powerless to stop it, which is what allows it to happen in the first place. Women in nontraditional jobs, often subjected to strong pressures to "go along to get along," are especially deprived of support in trying to stop workplace harassment. Women firefighters may come to believe that harassment is what they have to expect as a newcomer in a male-dominated workplace.

Sexual harassment is anything from blatant acts such as physical assault and *quid pro quo* pressures ("Sleep with me, or you won't get hired") to more subtle behavior such as persistent, unwanted requests for dates, displays of pornography in the workplace, and jokes that put women in subordinate, sexual roles or call attention to their gender.

Why sexual harassment is not reported

Even though it is illegal and a violation of most employer's policies, sexual harassment is only reported about five percent of the time. A woman who is harassed is much more likely to leave her job, request a transfer, or suffer in silence and hope the problem goes away. This is frustrating to everyone who wants to see sexual harassment in the workplace stopped.

---

*The vast majority of harassers--95 percent--are men, and most victims of harassment are women.

Verbal forms of harassment may include

- sexual jokes or teasing;
- suggestive comments or sounds;
- remarks about a person's clothing, body, or sexual activities (real or imaginary);
- pressure or demands for dates or sexual activities;
- derogatory name-calling or sexual references; and
- threats or intimidation.

Visual forms of harassment may include

- exposure to photographs, cartoons, drawings;
- gestures and other body language; and
- leering and other facial expressions.

Physical forms of harassment may include

- touching, patting, squeezing, pinching, brushing against, cornering; and
- physical aggression or intimidation.

Women firefighters who have been harassed and chose not to report it have described why they made this decision:

> I didn't want to be labeled a troublemaker, and I didn't feel a positive outcome was possible.

> Much of the harassment occurred when I was on probation and felt I could not speak out…I did attempt to speak to my officers; all of them shrugged off my appeals for help.

> I didn't want to be "singled out" even more.

> I want to keep my job. It's clear that those who seek legal recourse can't come back to work.

> I didn't want to get someone suspended or fired. I just wanted it to stop.[3]

There are many reasons women do not report sexual harassment. Some are complex and subtle, having to do with how women are brought up to view themselves and to behave. Others reflect the dynamics of a male-dominated workplace. Following are some of the most common reasons sexual harassment is not reported.

- Reporting the incident usually means an invasion of the victim's privacy. Harassment involves very personal interactions. Many women are uncomfortable with the prospect of having to discuss such subjects with their supervisors or other investigators, or of having to explain and relive the incidents. This is particularly true where the work environment in general is unsupportive, or where the victim has reason to believe the complaint process is not confidential. She also may fear, with good reason, that her own personal life or events in her past will be investigated, exposed, and presented as being somehow relevant to the harasser's behavior.

- Victims of harassment often believe that pursuing the matter would do no good and might even make matters worse. Often these concerns are justified. Women on many fire departments have little reason to believe management will do anything to stop the problem. A woman may endure years of unwelcome, abusive behavior without ever finding herself in a position where filing a complaint actually would solve the problem.

- Victims also legitimately fear retaliation: that the harasser will increase the harassment, or that coworkers or the employer will strike back in some way. The more the victim depends on her income--for example, if she is a single parent--and the fewer options she has for finding another job, the more likely she is to put up with harassment rather than jeopardize her paycheck by complaining.

- Because women are taught to be caretakers and nurturers, a woman may feel sorry for the harasser and not want to get him in trouble. She may try to find excuses for his behavior, or attempt to convince herself that something in her own behavior caused or contributed to the harassment.

- Victims of harassment may fear isolation and the loss of any friends or allies they have in the workplace if they rock the boat by complaining about the harassment. This is particularly likely in a workplace that fails to support those who report harassing behavior by filing a complaint.

Ideally, an employee who is harassed will confront the harasser, report the behavior to a supervisor, or seek support from counselors or from friends who have been harassed. But most victims of harassment feel disempowered and often think they have few options. This leads them to react in ways that neither stop the behavior nor let the employer know a problem exists: denial, trivializing, or excusing the behavior, or trying to appease the harasser in the hope that he/she will stop. If a harassment victim takes any action, it is usually either to request a transfer or to quit his/her job.

What constitutes illegal harassment

Federal law defines illegal sexual harassment as:

Unwelcome sexual advances, requests for sexual favors, and other verbal or physical conduct of a sexual nature when

(1) submission to such conduct is made either explicitly or implicitly a term or condition of an individual's employment,

(2) submission to or rejection of such conduct by an individual is used as the basis for employment decisions affecting such individual, or

(3) such conduct has the purpose or effect of unreasonably interfering with an individual's work performance or creating an intimidating, hostile, or offensive working environment.[4]

Helpful though it would be, it is impossible to make a comprehensive list of all behaviors that could constitute sexual harassment. Many situations must be examined individually; what is harassment in one case may not be in others. This can seem confusing until one focuses on the key concepts: "unwelcome," "intimidating or hostile," and "interfering with work performance." What is important is the **effect** of the questionable behavior on its victim. Behavior that is highly amusing to one person may be very unwelcome to another and therefore constitute harassment.

In cases that go to court, the court must determine whether the victim was offended by the behavior, and whether that reaction was "reasonable." Courts have differing ways of determining what is "reasonable." Some use a "reasonable person" standard: besides being actually offensive to the victim, the harasser's conduct must be such that it would affect the work performance and psychological well-being of a reasonable individual. But because a woman's perspective may differ substantially from a man's, some courts have adopted a "reasonable woman" standard instead.

A "reasonable person" standard does not consider the difference between women's and men's views of appropriate conduct. For example, in one study, 67 percent of men surveyed said they would be complimented if they were propositioned by a woman at work. When women were asked if they would take such a proposition from a man in the workplace as a compliment, only 17 percent said yes.[5]

Stopping sexual harassment

Most sexual harassment in the fire service continues because fire departments have not taken aggressive steps to stop it. If women did not have to fear for their jobs, their future promotability, or their effectiveness and safety at work when they report harassment, reporting would be much more common. If all fire departments had a zero-tolerance policy towards harassment that was enforced at all levels, harassment would be much less likely to occur in the first place. While it should not be the job of women firefighters to develop policy and encourage managers to comply with the law, this responsibility does often fall on their shoulders. Policy guidance is available in the U.S. Fire Administration publication, *Many Faces, One Purpose: A Manager's Handbook on Women in Firefighting*, and from Women in the Fire Service and many other working-women's organizations.

Notes:

[1] Woodland v. Houston, 1995 U.S. Dist. LEXIS 2749 (S.D. Tex.).

[2] Women in the Fire Service, Inc., unpublished survey data, 1995. Earlier surveys by Women in the Fire Service (1990), by Diane Sanchez of Sunset Associates (1991) and Rosell, et al. (1995) found from 58 percent to 75 percent of women firefighters had experienced sexual harassment on the job. [Rosell, Ellen; K. Miller and K. Barber. "Firefighting Women and Sexual Harassment," Public Personnel Management, Vol. 24 no.3 (Fall 1995): pp. 339-350.]

[3] Women in the Fire Service survey, 1990.

[4] EEOC Guidelines, 29 C.F.R. §1604.11.

[5] Goleman, D. "Sexual Harassment: It's About Power, Not Lust" New York Times, October 22, 1991; C1, C12.

If you are sexually harassed

Fire department employees or volunteers who are the victims of sexual harassment should

- Know the department's policies prohibiting sex discrimination and sexual harassment, and the procedures for filing a complaint.

- Politely and firmly tell the harasser to stop, in front of witnesses if possible. If you are unable to confront the harasser directly, write a letter and give it to the harasser in the presence of a witness. Keep a copy of the letter in a safe place, not at work. Or speak to your supervisor, to Personnel/Human Resources, or to an EEO officer.

- If the harasser repeats the conduct, inform your supervisor immediately and follow it up with a note or letter to the supervisor. Again, keep a copy in a safe place, not at work. It is important that you give notice of this offensive behavior to your employer and that you have a record that you did so. (Courts generally will hold an employer liable for sexual harassment only where it can be shown that the employer knew, or should have known, about the conduct.)

- Document all incidents in a diary with times and places, names of witnesses, what was said or done, and an exact account of your response and any physical or emotional stress you experienced. Make sure your notes are accurate and focused on current and relevant concerns: if they are later used in a lawsuit, they might not be considered confidential. Keep this log in a safe place, not at work. Do not destroy your original notes: if you need to copy them over for clarity, do so, but be sure to keep the originals. (Continued on page 37)

If you are sexually harassed (cont.)

- If the behavior involved criminal conduct, such as rape, attempted rape, battery or sexual assault, file a police report. Failure to report the crime may be used against you later if you decide to file a lawsuit.

- Talk with other women on the job or who have left the job, especially those who have worked with the harasser in the past; they may have been victims of harassment also. Do not be surprised or discouraged if other women do not support your decision to fight or report the harassment.

- Keep records of all of your evaluations, promotions, and other parts of your employment record, in the event that your complaint results in retaliation against you.

- Contact your union, labor organization, or employee group for assistance. If you work under a contract, be familiar with its anti-harassment clauses, its provision for arbitration of grievances, and your rights under the bylaws of your union.

- Contact women's organizations for support. You are not the first woman to be victimized by this kind of behavior. Some chapters of the National Organization for Women (NOW) sponsor support groups for victims of harassment, and many chapters of a group called "9 to 5" can give you support as well. Other women's professional and trade groups, such as Women in the Fire Service, also may be able to offer advice and resources.

- Be aware that both sexual harassment and the decision to take action against it are very stressful. If your employer offers an employee assistance program whose confidentiality you trust, use it. Try not to take the blame onto yourself. What happened to you was not your fault, the harasser did not "mean well," and he was not doing it because he likes you. You do not have to tolerate sexual harassment as the price of being a firefighter, and you do not have to adapt to the ways of your coworkers at all costs. If you have friends and coworkers who can offer support, depend on them. Be aware, however, that the general public is largely uneducated and often unsympathetic on this issue, particularly for women in nontraditional jobs.

- If informal or internal remedies fail to stop the harassment, if you lack confidence in the internal remedies (for example, if the person in charge of investigating your complaint is the one who is harassing you, or has threatened retaliation), or if you are retaliated against for complaining about the harassment, you should consult an attorney. Look for a lawyer with experience in handling employment discrimination complaints. Try to obtain private counsel, even though local, State, or Federal human rights agencies may assign a staff attorney to handle your formal complaint. The women's bar association or working women's advocacy groups may be able to refer you to an experienced attorney.

- File a formal complaint with the Federal Equal Employment Opportunity Commission or with your State or local fair employment practices (FEP) agency. There are specific time limits for filing such complaints, depending on whether the harassment is continuing or retaliation occurred, and whether your State or locality has an FEP agency. Title VII requires that you file with the EEOC within 180 days of the last act of discrimination. In areas with State or local FEP agencies, that time period is extended by 90 days if you filed with the local agency first; some States require you to file with the local agency first. To protect yourself, file as soon as possible. You may wish to file the charge and then ask that it be held while you attempt to resolve the problem internally. You can always drop the charge if matters are resolved to your satisfaction.

Firefighter marriages and fire department policy

Fire departments historically have had a strong family tradition. Sons have followed fathers into fire service careers for generations. Brothers have served side-by-side as career or volunteer firefighters. Now that women are an integral part of the fire service, a new family tradition has emerged. Although women may follow their fathers (or mothers) into the fire service, and sisters and brothers may both choose firefighting vocations, the most common type of family relationship for women with a coworker is that of marriage.

According to a 1995 survey by Women in the Fire Service, 32 percent of women firefighters were married to or in long-term relationships with other firefighters. Nearly three-fourths of the relationships were between firefighters on the same department, and most had developed after both employees were hired.

It should not be surprising that these figures are so high. Once a person has finished school, the workplace is the most common place to meet friends, social partners, or mates. This is true in any profession, but seems especially so with firefighters. The nature of the job, particularly the unusual hours and work environment, can make a conventional social life difficult. Strong bonds of friendship and loyalty have always developed among firefighters who work closely together under difficult circumstances. It is natural that these same feelings would develop between men and women firefighters, and that good friendships might potentially lead to further involvement or commitment.

Fire departments have reacted to this trend in firefighter relationships in widely varying ways. Some have taken a completely laissez-faire attitude and have not interfered in any way with firefighter couples, even allowing them to continue working in the same station together. Other departments have taken the opposite approach, attempting to impose very restrictive policies on these employees. An extreme case of this type of policy was one that prohibited marriages between employees and made marriage to another firefighter cause for dismissal. The policy was made effective retroactively so that the one woman on the department (who was married to a coworker) would be affected. Another example is of departments that attempt to discover which couples are dating or otherwise emotionally involved with each other, and reassign one or both parties. Such policies go beyond legitimate employer concerns into the realm of harassment.

Policies against nepotism (favoritism to a relative) are a legacy of 19th-century political "spoils" systems and were originally put in place to prevent local political officials from appointing their relatives to jobs. The policies are more common in some parts of the country than others; of fire departments recently surveyed, less than a third had a formal anti-nepotism policy. Some policies include strong statements such as the following:

> The employment of relatives in the same organization tends to have a number of undesirable results. In the interest of preventing potential abuses in hiring, supervisory authority, or the appearance thereof, it is the policy of the City to limit hiring and supervision of relatives by City employees.

Others take a more moderate approach:

> The department recognizes that there are many situations where two individuals who have a personal relationship may appropriately be allowed to work in the same program, activity, or location without adverse impact. However, under circumstances where work, safety, morale, or impartial supervision is demonstrably and adversely impacted by a personal relationship, the affected employees may be accommodated by the reassignment of one or the other.

The anti-nepotism policies of some cities restrict the hiring of relatives but do not address the potential for employees to become related through marriage once both individuals are employed. Faced with policies that impose employment restrictions on relatives, firefighters have felt forced to hide relationships or even lie about them. Since some nepotism policies seem to punish people for marrying, couples may choose to live together without legal bonds, although some departments have tried to apply strict nepotism policies to unmarried department members who live together, as well as to married couples.

Working to change anti-nepotism policies

Firefighters seeking to change their department's position might ask management to consider the goals and effects of nepotism policies. Are the policies written to address real problems? What specific problems have actually arisen in the department from the employment of relatives or couples? What other, less restrictive options are available for dealing with these problems? Are policies applied evenhandedly, or are couples ostracized while other relatives are left alone?

When restrictive nepotism policies are applied to married couples in the fire service, women firefighters are affected disproportionately. For example, on a department of 300 members with 10 women firefighters, three of those women are statistically likely to be involved with other firefighters. Although their partners would represent only one percent of male firefighters, the women involved represent 30 percent of all women on the job. If a department enacts a restrictive policy limiting employees' career opportunities because of their relationships, that department is accepting that one in three of its women firefighters will be limited in this way.

The original purpose of nepotism policies was to ensure that employees or applicants would be judged by their own merit and not by who they were related to. Just as employees should not have an unfair advantage because of their personal relationships, neither should they be unfairly disadvantaged on the same basis. In times of downsizing, few organizations can afford to throw away an individual's contribution. Yet this can be the effect when restrictive nepotism policies are applied when no actual problems exist.

Relationships in the workplace are a fact of life. They need not be viewed as a problem to be solved or as a situation to be prevented. Where real problems do exist, either with the couple themselves or due to coworkers' or managers' resentment of the couple, they should be dealt with individually. It is possible to develop policies that respect the rights and privacy of individuals and at the same time to maintain a professional work environment for all employees.

Firefighter reproductive safety issues

Male and female firefighters face hazards to their reproductive health in the course of their work. These hazards include such common fireground exposures as heat and carbon monoxide, as well as a wide array of less frequently encountered risks. Exposure to such hazards may make it more difficult to conceive a child or less likely that the fetus will be carried to term. Exposure may affect fetal development and possibly cause birth defects. Toxins may be transmitted through the bloodstream to the fetus, or to the infant during breastfeeding.

Concern over this issue has arisen in a roundabout way. It wasn't long after the first women became career firefighters before some fire chiefs voiced their awareness of the reproductive risks of firefighting, but this usually was expressed less as a concern for worker safety than as an argument against women being firefighters at all. Rather than taking steps to address the risk, the fire service saw the ability to get pregnant as further evidence that women didn't belong on the job.

Women firefighters in those early years rarely had the option of planning a family. Many felt that, in being hired as a firefighter, they had tacitly agreed not to get pregnant. Women who did get pregnant sometimes faced the bleak prospect of losing their jobs, and were almost always in the uncomfortable position of not knowing how their pregnancy would affect their employment or their income. Fire departments have been slow to address these issues, and it has only been in the 1990's that fire service leaders have begun to take an interest in safeguarding the reproductive health and safety of all personnel.[1]

In addition to protection from risk, working parents of both sexes need time off work when their child is born, adopted, or seriously ill. The majority of employees will need this kind of leave at some time during their careers. Recent changes in U.S. law have set a minimum standard for family leave; firefighters may wish to work for departmental policies that go beyond this minimum.

Firefighting and reproductive health

The research on firefighting and reproductive safety is too sparse to support solid conclusions; much more research is needed on firefighters' exposures to heat, noise, and toxic chemicals, and on the impact of these exposures on reproductive health. The information that is available, however, clearly indicates the presence of significant risks.

One of the most common products of combustion is carbon monoxide, to which the developing fetus is particularly vulnerable. Fetal hemoglobin has a higher affinity for carbon monoxide than does the mother's hemoglobin. This means the fetus will be affected more than the mother; exposures that produce only moderate symptoms in the mother can be fatal to the fetus.[2]

A father's employment as a firefighter may result in a higher risk of congenital heart defects for his children. A study in British Columbia in the late 1980's found a link between increased rates of atrial and ventricular septal defects in the children of male firefighters. The study concluded,

> The occupation of fire fighting merits further attention in view of increasing knowledge regarding male-mediated teratogenesis (birth defects).[3]

In 1989, researchers at Johns Hopkins University gathered reproductive data from firefighters. Their research found that chemical exposures pose risks to the reproductive health of both male and female firefighters, while carbon monoxide poses risks to the developing embryo.[4] High ambient heat in the firefighter's working environment also causes problems. The core body temperature of firefighters involved in interior operations at structure fires can rise high enough to impair sperm production in male firefighters and pose a risk of birth defects to an embryo carried by a pregnant firefighter. [Another medical authority has linked maternal hyperthermia early in pregnancy to neural tube defects in the fetus.][5] Maternal exposure to high levels of noise--firefighters' exposures include sirens, air horns, vehicle engines, and power tools--is linked to decreased fetal weight and increased fetal mortality.[6]

Preliminary results from a 1995 survey conducted by Women in the Fire Service showed an incidence of birth defects, premature births, and other childbirth problems in the children of women firefighters that may be higher than normal. The incidence was significantly higher when the child's father was also a firefighter.[7]

One group of EMS workers found high rates of gynecological problems in women EMT's and paramedics who worked with 800mHz radios or in close proximity to video display terminals in their ambulances. Menstrual irregularities, miscarriages, birth defects, uterine cysts, cervical abnormalities, or cancer were found in as many as 100 of the service's 680 women workers. According to the head of a group investigating the problem, when the terminals and radios were first installed, 15 women at one station alone developed menstrual irregularities.[8]

Glossary

Hemoglobin: the chemical compound in the blood that carries oxygen to the tissues.

Hyperthermia: high body temperature.

Teratogenetic: causing birth defects.

For the nursing mother, the risk continues after the child is born. Studies on cigarette smoking and other chemical exposures make it appear likely that the same or similar chemicals in the fire environment are passed on to the infant through maternal milk.[9]

Fetal and maternal susceptibility to toxins in the workplace have been used often as an excuse to keep women out of particular jobs. Once it is shown that men, too, are harmed by these toxins (vinyl chloride and lead are two examples), the toxins typically are regarded in a new light: they become "workplace hazards." The prevailing view quickly alters to find ways to manage exposure risks rather than prevent employees from working. The same evolution of attitude and policy probably will hold true for the fire service. According to one of the Johns Hopkins researchers,

For the few reproductive toxins that have been well studied, evidence demonstrates effects mediated through both males and females. In fact, one author has suggested that males may be more sensitive to exposure to reproductive toxins.[10]

The legal background

Two Federal laws and two court decisions provide the framework for reproductive safety and family leave policies. The first law is the Pregnancy Discrimination Act of 1978 (PDA), an amendment to Title VII of the Civil Rights Act of 1964. The PDA broadens the definition of sex discrimination to include discrimination based on pregnancy and childbirth. It states

Women affected by pregnancy, childbirth or related medical conditions shall be treated the same for all employment related purposes... as other persons not so affected but similar in their ability or inability to work...[11]

This law applies to all employers with 15 or more employees, and to employment agencies and labor organizations.

The PDA guarantees pregnant women access to benefits already in place for other workers. For example, company health insurance plans may not exclude coverage for pregnancy and childbirth. An employer may not refuse to hire or promote a woman solely on the basis of pregnancy. Disability caused by pregnancy must be covered under an existing disability program. If the employer gives temporarily disabled workers light duty, it must also give pregnant employees light duty. Most importantly, a woman cannot be fired from her job arbitrarily if she becomes pregnant, nor can she be required to take an extended leave that is not medically necessary.[12]

In the years following the passage of the PDA, several States developed policies mandating certain types of benefits for pregnancy and childbirth. California passed a law that guaranteed up to 4 months of unpaid leave for childbirth and recovery. The California Federal Savings and Loan Association challenged this law as a violation of the Pregnancy Discrimination Act, arguing that since the PDA required all employees to be treated the same, special benefits for pregnancy and childbirth were illegal.

The case went to the Supreme Court, and in 1987 the Court upheld California's policy and clarified the intent of the Pregnancy Discrimination Act. The majority opinion stated that the Federal law was intended to "construct a floor beneath which pregnancy benefits may not drop, not a ceiling above which they may not rise."[13] This decision allowed individual employers to develop policies specifically for maternity.

Some fire departments, as well as employers in other potentially hazardous work environments, then developed policies for employee pregnancy based on the concept of "fetal protection." Hazards in many workplaces may harm a developing fetus or the reproductive health of men or women workers. Policies based on fetal protection often required a woman to leave a certain type of job, such as active firefighting, at some point in her pregnancy; in many cases, as soon as the pregnancy was known. This requirement was based not on the woman's ability to perform her job, but rather on the employer's concern that some aspect of her work might prove harmful to the developing fetus.

The Supreme Court addressed fetal protection policies in its 1991 decision in *UAW v. Johnson Controls*. This case concerned a battery manufacturer that had barred all fertile women from working in jobs using lead because lead could harm a fetus. The Supreme Court rejected this policy as a form of sex discrimination and a violation of the PDA. The majority opinion stated

Decisions about the welfare of future children must be left to the parents who conceive, bear, support, and raise them rather than the employers who hire those parents... It is no more appropriate for the courts than it is for individual employers to decide whether a woman's reproductive role is more important to herself and her family than her economic role. Congress has left this choice to the woman as hers to make.[14]

The second Federal law is the Family and Medical Leave Act (FMLA) of 1993, which guarantees 12 weeks per year of unpaid leave to employees in order to care for a newborn, newly adopted, or seriously ill child. The employer must continue health care benefits during this leave, and must reinstate the employee into his or her original position, or into an equivalent position with equivalent pay and benefits. Employers may require employees to use any paid sick leave and vacation time as part or all of the 12 weeks of FMLA leave. (*For more information on the Family and Medical Leave Act, see pp. 48-50.*)

Parental leave is not a new concept. The parental leave policies of six European countries (Austria, Germany, France, Italy, Finland, and Sweden) guarantee from 12 to 52 weeks of leave, with 60 to 100 percent salary retained during all or part of the leave. Canada offers 17 to 41 weeks of parental leave, with 15 weeks at 60 percent salary guaranteed. By contrast, the 12 weeks of leave mandated by the FMLA are entirely unpaid.

Developing policies for fire departments

Policies may be developed in a number of ways. Fire department management or city administrators may draft a policy after consulting legal counsel and employee representatives. Where neither the fire department nor the city will implement a suitable policy, the matter is often negotiated by the firefighter's union. This somewhat distorts the purpose of the policy, as it converts workplace safety into a benefit and subjects it to the tradeoffs of the bargaining table. One advantage, however, is that male-dominated union locals are more likely to push for reproductive safety policies that benefit all firefighters, rather than ones that deal only with pregnancy.

Although employment policies that relate to reproductive safety are often categorized as "maternity" policies, three separate areas of risk or need should be addressed:

1. Reducing reproductive risk for the employee who is pregnant, breast-feeding, or attempting to conceive a child.

2. Providing adequate leave for the woman during the time she is disabled as a result of childbirth.

3. Providing adequate leave for new parents surrounding the birth or adoption of a child.

Reducing risks to reproductive safety. The general trend among career-level fire departments is to guarantee alternate, nonhazardous duty to a firefighter during the term of her pregnancy. All women workers need to be able to take leave from their jobs for the time surrounding childbirth, but women firefighters' needs are more complex. While the PDA guarantees pregnant women the right to stay on the job as long as they are able to perform, there is real concern as to whether women firefighters should continue in their usual job assignments while pregnant.

Many physicians agree that women should stop fighting fires and doing other high-risk work at some point in their pregnancies. Exactly what point is most appropriate is a subject of some debate. Because many environmental hazards are most dangerous to a fetus during the first trimester (the first 3 months a woman is pregnant), many fire departments want women to leave hazardous duty as soon as their pregnancies are known. The majority of women prefer this arrangement as well.[15] Nonetheless, the Supreme Court decision in *Johnson Controls* strongly suggests that such transfers cannot be mandatory.

As of 1996, the law upholds a woman's right to continue working in a hazardous environment even during pregnancy. If faced with a choice between a higher-risk pregnancy and the economic devastation caused by a significant loss of pay and benefits during pregnancy because no alternate duty is offered, many women will feel compelled to choose the former. The International Association of Fire Fighters (IAFF) in 1992 adopted the position that the pregnant firefighter should be offered the opportunity for voluntary transfer from firefighting at any time during her pregnancy without loss of pay or benefits.

As relevant research increases, more attention is being given to the reproductive risks of parental exposure near the time of conception. Women and men who are trying to conceive should avoid environments that may endanger their reproductive health. Thus, nonhazardous duty should be offered to any employee attempting to conceive a child. Because of the potential for toxins to be transmitted to an infant through breast milk, nursing mothers also should have the option of alternate duty.

Alternate, nonhazardous duty should be meaningful work that does not penalize the employee in any way. Firefighters on alternate duty have worked productively in such areas as training, public education, prevention and inspection, policy development, and communications. Fire departments of all sizes have found ways to use these employees productively in roles that do not present reproductive hazards; the department and the employee both can benefit from the employee's short-term assignment to a noncombat position.

> Pregnant employees may be offered alternate duty even if that option is not available to other employees. The legality of this type of variance in treatment was clearly upheld in *California Federal Savings & Loan*. Having a more generous policy for maternity than for other types of disability does not constitute illegal sex discrimination.

Nonhazardous assignments for workers seeking to avoid reproductive risks are not necessarily the same as "light" duty that may be made available to those returning from an injury. For example, light duty might require a firefighter to fuel the department's staff cars every day, but excessive exposure to gasoline and diesel fumes during pregnancy may cause health problems. Similarly, doing arson investigation in freshly burned buildings is probably not the best choice of assignment for a firefighter who needs to avoid reproductive risks.

Firefighters on nonhazardous duty should be allowed to take training or recertification classes that other firefighters are undergoing, as long as the classes do not involve risks. This not only keeps them from falling behind in their training and saves the department from having to train them later, but also allows them to remain in touch with coworkers and with suppression operations while they are reassigned.

A fire department is not required to offer alternate-duty assignments to pregnant firefighters unless it has a policy of reassigning all employees who have temporary disabilities. Pregnant employees may, however, be offered alternate duty even if that option is not available to other employees. The legality of this type of variance in treatment was clearly upheld in *California Federal Savings & Loan*. Having a more generous policy for maternity than for other types of disability does not constitute illegal sex discrimination.

Maternity leave. Only a generation ago, motherhood and work outside the home were considered mutually exclusive. If a woman had a job at all, she was expected to quit once she started having children. Many employers refused to deal with maternity as an issue in the workplace. Employers were allowed to treat pregnant employees in any way they chose, and the law offered no protection for these women. As recently as 1964, 40 percent of all employers terminated workers who became pregnant.

Even after women had been in the career fire service for years, many of their employers had not developed policies to address pregnancy. A 1995 survey showed that 60 percent of fire departments that employed women had no written maternity policy, and an additional 23 percent had only a city- or countywide policy that did not specifically address the needs of women in hazardous professions. Only 30 percent of fire departments employing women had some sort of written maternity policy specifically for firefighters, and in many cases this simply consisted of language specifying that pregnant women could use their sick leave and vacation time, or (as Federal law requires) that departmental provisions for employees injured off duty applied to pregnant firefighters.[16]

Maternity leave policies address the period of time when a woman is physically unable to work as a result of pregnancy and childbirth. All women who go through childbirth are temporarily disabled by it to some degree, whether for a few days or for several months. Every workplace should have a policy that allows women to have children and physically recover afterwards. When a woman's pregnancy does not restrict her ability to do her job, and when she suffers no complications from the pregnancy or delivery, a typical maternity leave might be six to eight weeks. Extensions should be available for cases where the pregnancy or delivery are very difficult. Employers sometimes provide this time as paid or partly-paid leave specifically for childbirth; in other instances, women must use sick or vacation time to maintain their income.

In a minority of cases, a pregnant employee will be unable to work even in an alternate-duty assignment because of health complications during or after pregnancy. A leave of absence during pregnancy and childbirth (beyond the 12 weeks mandated by FMLA) should be available that will accommodate difficult pregnancies and recovery from a delivery that involves complications. Such leaves are usually unpaid, but they should include continued health insurance coverage.

Many fire department policies require the opinion of a doctor, often the employee's personal physician, to determine how long a pregnant woman can work safely in her fire service job. Since many physicians are not familiar with the actual demands and hazards of firefighting, they should be educated about the job before they are asked to give such an opinion. Fire departments may wish to develop a standard physician's release form for this purpose that specifically lists the requirements of the job. Any such form used in cases of pregnancy also must be used in a comparable way for other nonduty-related disabilities, such as off-the-job injuries.

Parental Leave. Most women will be physically capable of returning to full firefighting duty within six to eight weeks following the birth of their babies. However, a new mother or father may wish to spend more time with an infant beyond the time needed for physical recovery. Prodded by State and Federal law, employers now offer leave to employees who want or need more time to be full-time parents to a new baby or a newly adopted child.

Many employers, including city or county governments and fire departments, have recognized parental leave as a policy that can benefit both workers and their employers. Parental leave policies that surpass the requirements of the FMLA and State laws may be negotiated into contracts, developed by managers as standard practices, or enacted as local ordinances. Four months is a common length for unpaid parental leave in the U.S., although some employers provide up to a year. In an increasing percentage of cases, such leave also may be used, like FMLA leave, to care for an ill or bedridden child or older relative.

Educating employees

The fire department should provide education as part of its reproductive safety policy. All employees must understand the hazards firefighting poses to their reproductive health. A qualified physician or other professional who is well versed in the existing research on the issue should conduct classes on this subject for all firefighters and officers. Women firefighters also should be educated early in their careers about the options that exist for them if they become pregnant.

Conclusion

Reproductive hazards and family needs pose manageable challenges to fire departments, just as they do to other employers. Fire departments that continue to view reproductive safety from the narrow perspective of pregnancy (i.e., as a "women's problem") do a disservice to their employees. And employers that do not implement policies addressing employee pregnancy adequately cannot be said to have truly accepted women's presence in the workplace.

Many fire departments have arrived at good solutions that balance the needs of the employee and the employer. A policy that allows the employee to continue as a contributing member of the department during pregnancy, and that assures her of continued pay, benefits, and seniority, is an attainable goal for fire departments of all sizes.

Notes:

[1] Willing, Linda, "Maternity Survey." *Women in the Fire Service Quarterly*, Summer 1987, pp.1-6.

[2] Marzella, Louis, M.D., Ph.D., et al. "Carbon Monoxide Poisoning." *Practical Therapeutics*. Vol. 34, no. 5, pp. 186-194.

[3] Olshan, Andrew F., et al. "Birth Defects Among Offspring of Firemen." *American Journal of Epidemiology*, Vol. 131 no. 2, p. 312.

[4] McDiarmid, Melissa, M.D., et al. "Reproductive Hazards of Firefighting I and II." *American Journal of Industrial Medicine*, 19: 433-472 (1991).

[5] Milunsky, A., et al. "Maternal heat exposure and neural tube defects." *Journal of the American Medical Association*, 268:882-885,1992.

[6] McDiarmid, *op. cit.*

[7] Women in the Fire Service, Inc., unpublished survey data, 1995.

[8] Floren, Terese M. "Health Problems Raise Questions for NYC EMT's." *Firework*, April 1993, p. 1.

[9] "Pollutants in Human Breast Milk." *Human Toxic Chemical Exposure*, undated.

[10] McDiarmid, op cit.

[11] 42 U.S.C. §2000e (K).

[12] *EEOC Questions and Answers on Pregnancy Discrimination*, C.C.H. ¶3951 (April 20,1979).

[13] *California Federal Savings & Loan v. Guerra*, 107 S. Ct. 683,42 F.E.P. 1073 (1987).

[14] Supreme Court of the United States, No. 89-1215. *UAW v. Johnson Controls Inc.*, majority Court opinion delivered by Justice Blackmun on March 20, 1991.

[15] "Those surveyed were asked to describe...the ideal maternity policy. The vast majority outlined a policy that would provide light duty for the term of the pregnancy...then leave permitted during actual delivery and for three to six months after birth." Willing, Linda, "Maternity Survey," *Women in the Fire Service Quarterly*, Summer 1987, p. 4.

[16] Women in the Fire Service, Inc., unpublished survey data, 1995.

Family and Medical Leave Act

The FMLA of 1993 expanded sick leave and family-care leave benefits for millions of U.S. workers. It guarantees employees up to 12 weeks a year of unpaid leave for the birth or adoption of a child or the placement of a foster child, or for a serious health condition of an employee or his/her spouse, parent, or child.

Many States already had family leave laws on the books before 1993. The FMLA is notable not only because it extended this benefit to workers in States that did not have such provisions, but because it requires that health care coverage of employees be continued during such leaves. Most State requirements do not have this provision.

The FMLA provides minimum guarantees and does not take away benefits provided through employer policy or collective-bargaining agreements. Thus, if an employer already is required, by contract or State law, to provide more family and medical leave than the FMLA mandates, the FMLA does not reduce that requirement. Employers must comply with whichever provisions are most generous to the employee. Correspondingly, collective bargaining agreements may not be used to diminish workers' rights under the FMLA.

The FMLA applies to all public-sector employers subject to the Fair Labor Standards Act and federal minimum-wage laws, and to all private employers with 50 or more employees. Employees must have worked for the employer for at least a year, and must work an average of 25 or more hours a week, to be covered.

Key provisions of the Family and Medical Leave Act:

- Workers may take up to 12 work-weeks of unpaid leave during any 12-month period in order to care for a newborn child, a newly adopted child, or a newly placed foster child; to care for a spouse, child, or parent with a serious health condition; or due to a serious health condition that leaves the employee unable to work.

- The employer must continue the employee's health care benefits during FMLA leave. The employer must maintain employee coverage under a group health plan if the employee would have been eligible for such coverage if he or she had not been on leave. (The law does not address the issues of other benefits such as pension contributions during leaves.)

- If employees normally pay a portion of health plan costs, they must continue to pay this portion during leaves. If an employee chooses not to return to his/her job after a leave, the employer can, under certain conditions, demand repayment of the health care premiums it has paid on behalf of the employee.

- When an employee returns from leave, the employer must put her or him back into the position previously held, or into an equivalent position with equivalent pay and benefits. The employee must receive any unconditional pay increases that occurred during the leave, as well as certain other pay bonuses. "Key employees," defined as the highest-paid 10 percent of the workforce, may be denied reinstatement in certain cases if this would cause substantial economic harm to the employer.

- Leave taken for the birth or placement of a child may be taken any time in the 12 months that follow the date of birth or placement.

- "Intermittent" or "reduced-schedule" leaves are permitted for birth or placement leaves; they must be made available for health leaves. This means that if a worker has a medical need for small chunks of leave--for example, to take a family member to and from the hospital for weekly medical treatments--or needs to work on a reduced schedule (such as one that allows him or her to be home at night), the employer must make these leaves available. The employer may temporarily transfer employees using these leaves to a different position--for example, from a shift schedule to a 40-hour job--to better accommodate the needed leave, as long as the new position has equivalent pay and benefits. Employers are not required to allow intermittent or reduced-schedule leaves for the birth or placement of a child, but they may agree to do so.

- Employers may require employees to use their paid vacation and sick leave as part or all of the 12 weeks of leave mandated by FMLA. Compensatory ("comp") time given in exchange for Fair Labor Standards Act overtime compensation is not considered a form of accrued paid leave. An employer, therefore, cannot require an employee to use comp time as part of an FMLA leave, although employees may do so at their own discretion.

- Employers do not have to allow workers to accrue seniority while on leave. FMLA leave does not constitute a break in service for purposes of pension vesting.

- Spouses who work for the same employer may be limited to a combined 12 weeks of leave in the case of childbirth, adoption/placement, or family illness (of someone other than the employee).

- Employees may be required to provide 30 days' notice of their intent to take leave for childbirth or child placement, or for other foreseeable reasons.

- The employer may require a doctor's certificate to verify the need for leave taken for health reasons. The employer may, at its own expense, require a second opinion and, if the two conflict, a final and binding third opinion. Employers may also require periodic recertifications and reports on the employee's status.

- Employees are not eligible for unemployment benefits during FMLA leaves.

- An employer may not restrict the number of employees who may be on FMLA leave at one time.

Definitions:

"Serious health condition" means an illness, injury, impairment, or physical or mental condition involving inpatient care (in a hospital, hospice, or residential health-care facility), or continuing treatment by a health care provider. It also includes treatments for various diseases and for surgery to repair injuries. Short-term illnesses are not covered, and common illnesses such as colds and flu are specifically excepted. Employees using leave for their own health conditions must be unable to perform their job functions, though employers must also approve intermittent leaves for employees to receive necessary treatments for early stages of diseases such as cancer.

A "chronic serious health condition" is one that requires periodic medical treatment and continues for an extended period, either continuously or as a series of episodes. Employees who suffer from chronic serious health conditions, or whose family members do, are eligible for FMLA leave, even if they are incapacitated for fewer than three days and do not visit a health care provider.

"Son or daughter" means a biological child, adopted or foster child, stepchild, legal ward, or other child under 18 for which the employee stands in the place of a parent. Sons and daughters 18 or older are included if they are incapable of caring for themselves due to a mental or physical disability.

"Parent" is defined correspondingly; the FMLA does not, however, mandate leaves to care for a nonmarital partner or for the parents of one's spouse. (Some State laws do mandate leaves for these family members.)

"Care" given by the employee need not be physical, but may refer to psychological care in the home as well as during inpatient stays in health-care facilities. The definition of "health-care provider" was expanded in 1995 to include social workers as well as any providers recognized by the employer. The employer's health-care provider is permitted to contact the employee's health-care provider in order to clarify information in the medical certification, but such contacts may not include a request for additional information about the employee's condition.

"Equivalent position" does not mean that the employee's new position is merely comparable or similar to the one she or he left. The job duties and all terms, conditions, and privileges of employment must correspond to those of the original position.

Employers are required to post notices in the workplace explaining the provisions of the FMLA. Most employers will want to go farther, including information about FMLA leave in employee handbooks or other material about benefit programs. Employers who violate the FMLA are liable for the wages and benefits that the employee loses, or for any actual monetary losses. Employers who have not acted in good faith are liable for double damages.

For more information, see the Family and Medical Leave Act, 29 U.S.C.A. §§2601-2654 (1993) or the Final Rules, Federal Register, January 6, 1995 , [60 Fed. Reg. 2180-2279 (1995)].

Child care and parenting issues for firefighters

Firefighters are in a unique situation when it comes to child care. Their children usually need nonparental care for more than 24 hours at a time. In addition, many firefighters are subject to emergency call-back during major incidents, and may need someone to take care of their children on a moment's notice at any hour of the day or night. These circumstances especially affect women firefighters, two-firefighter couples, and single parents. More than a third (36 percent) of women firefighters have children living at home, most with no other parent at home to provide child care. Almost as many women (32 percent) are married to or involved with other firefighters; 43 percent of these couples have children at home. One out of every 12 women firefighters is a single parent. Clearly, these issues have a significant impact on the ability of women to work as firefighters.[1]

Many cities and private employers have begun to take an interest in the child care needs of their employees. They recognize that child care problems cause absenteeism, reduced productivity and poor morale, and may lead to the loss of good employees. In particular, they know difficulties with child care are a significant reason women may not enter the fire service, may leave early in their careers, or may not return after having children.

Although employers are beginning to recognize this issue, most existing programs address only the needs of employees who work relatively conventional hours. The few employers who support child care centers for their workers most often maintain these centers only during extended business hours.[2] Centers that can accommodate 24-hour child care or emergency drop-ins at any hour are virtually nonexistent. Innovative management practices that may help some employees with child care problems, such as job-sharing and flexible hours, are either not feasible or not often made available for firefighters.

Creative solutions are needed and have begun to emerge. One fire chief has suggested that his city develop old fire stations into around-the-clock child care centers specifically for the benefit of firefighters' children. "We found that some qualified people do not apply for the job because they're concerned about what they'd do with their kids," he said. "I don't think it benefits the department if qualified people get away because of that."[3] In Great Britain, the London Fire Brigade provides child care allowances to parents who pay babysitters, and subsidizes spots in child care centers for its employees' children.[4] The Suisun City, California, Fire Department has increased the off-duty response of its full-time personnel to major incidents by outfitting and staffing one of its rehab vehicles to handle child care. Firefighters responding to the call can drop off their children at the fire station or at staging, to be cared for by members of the rehab team for the duration of the incident.[5] Other fire departments provide child care during off-duty training sessions, to encourage greater participation. Positive examples also are being set by some hospitals and airlines, which have many employees who need child care at unusual hours.

Being a firefighter and a single parent

In 1994, Women in the Fire Service gathered information from women firefighters who were single parents. Their comments are compiled here.

Coworkers responses to a woman becoming a single mother due to a divorce or break-up:

"My co-workers viewed my being a single parent mostly positively. I believe the fact that a lot of my time was taken up with being a parent made me less of a threat and more respectable, even though I was not married. They were often supportive and sympathetic."

"My co-workers are supportive. They realize the sacrifices it takes to be the first woman and the first mother on the department. Many are uncertain what to do except offer encouragement, but they do know how hard I work to see that the children are cared for properly."

"The boys' father tried to interfere with their caretakers while I was at the station on duty. I had to leave work several times to handle the situation. My co-workers were supportive of me when this happened."

"I split up with my boyfriend while my child was very young. I was still on probation with the fire department, but I decided I'd been doing it all myself anyway, and it shouldn't be much harder without him. Some of the guys I work with seem to empathize with me, but most don't seem to take me, my job, or my family seriously."

Coworkers' responses to a single woman choosing to have a child:

"My co-workers were a little bit miffed. I was several months along before I even knew I was pregnant, and they thought I hadn't told them because I would have to go on light duty and might lose my position...After all this policy stuff was cleared up, they were still slow in responding; however, it's getting better (now that) they've seen her, held her, and been called 'Uncle _____.'"

"The men felt that I had planned things out carefully, and they thought it was great. The women tended to say things like 'A baby needs a father.' When I ran out of sick time, many more men than women donated to my leave pool."

"Mostly my co-workers were very supportive. During my pregnancy, two negative situations arose...A friend of mine told me that a crew on another shift spent an afternoon discussing the 'morality' of my decision to have a baby. I spoke with my supervisor; later, the officer of that crew called me to apologize...The other situation involved a rumor that a particular fire officer had sent me a turkey baster in the mail; this never occurred."

"Most of my co-workers have been very happy for me. Occasionally a male co-worker will ask why I did this alone without getting married. They are usually receptive to my honest response. The harder thing to deal with has been racism. My daughter is from Ethiopia and is black, and I am white. There have been some ugly racist comments...I usually react so strongly that they never say anything like that again in my presence."

Custody issues

Women firefighters sometimes face threats to their custody of their children, based on the accusation that their work as firefighters, or the fact that they work overnight shifts, makes them an unfit parent. Comments from women who had been through this battle:

"My custody of my children was challenged by my ex-husband. He threatened me with it for years, and we even went to mediation to discuss it. The mediator told him that he had no basis to charge me with being unfit due to my job and work schedule."

"I chose to become a single parent while separated from my husband, now my ex-husband. In this state, the husband is the legal stepfather even if he is not the baby's biological father. My ex-husband wanted my child, or so he said, until the courts were going to make him pay child support. Then we had to spend thousands of dollars, including a blood test, to prove that he wasn't the father. After that, the court decided that it was in the best interests of the child to have a court-appointed guardian... Because of the guardian, I almost had to transfer to a desk job. One of the main things that helped sway them to my side was that I have 21 or 22 days a month off work."

"I once had my daughter's dad threaten to take her away from me because 'What kind of mom can you be when you work 24 to 72 hours straight?' But the fact that I'm home with my daughter more than any other full-time working parent stopped him pretty quick."

'The biggest drawback is the mental stress a court battle for your children places on you. The second-largest burden is the legal expenses. You try to maintain a balance, and it is nearly impossible, because every time you look at the children, you think about how important they are to you and how you want them raised as they should be. You smile as you walk out the door, and then you worry for the next 24 hours and wonder if you'll survive all the turmoil."

Child care options

Not surprisingly, child care was the number one problem for single-parent firefighters. Most had used a variety of care providers. Usually, this was a family member (most often a mother, sometimes both parents, a sister, or the child's father), a friend or a live-in nanny. Less frequently used were daycare businesses and paid babysitters who didn't live with the family. Almost all of the women said they'd had to take sick leave or other forms of leave occasionally when child care arrangements fell through. Sometimes this could be done "out front" (that is, by calling in and saying they couldn't come to work because their nanny was sick); in other cases, the woman simply called in sick herself.

- **Nannies and au pairs**
  "Over the years, I have been through many babysitter changes, using family, friends, day-care businesses, my daughter's dad (this didn't last long!) and finally settled down with a live-in nanny. This was the ideal situation... she also helped clean the house, which made it nice to come home and not have to clean all day! The few problems with a live-in were, obviously, sharing our house with someone else, dealing with my daughter's disrespect for her nanny, and having to be "mom" to another person... Unfortunately, after she'd been here for eight months, she had to leave to go care for her sick grandmother, so I am now back to bouncing my daughter between family and friends while I desperately try to find another nanny."

  "Due to the cost factor, I had to opt for a live-in nanny. She's great with the kids and very flexible with regards to my hours if I have to work overtime at the last minute. The kids love her. The down side is the loss of privacy. It took me the better part of a year to get used to someone else living in my home and sharing my space. But this is a relatively minor inconvenience... The biggest problem is that I don't make enough money on my own to pay for a nanny and still make ends meet. My ex has been very good with support payments, but if he didn't pay, I would have to take a second job."

  "My first au pair was a great success and stayed with me for 18 months. After that, I had a succession of different people over the years who looked after my child, some organized only hours before I had to go to work."

- **Family members**
  "I was very fortunate to have my parents available to babysit while I was on duty, and I am very thankful to them for their support. My children's father was no help at all."

  "Initially I called on neighbors and my mother. Thank heaven for moms!"

  "When I'm on shift, I take my daughter to day care. Her father picks her up and takes her to his house when he gets off work, and brings her to my station in the morning on his way to work. He also pays for the day care."

"My mother takes care of the baby while I'm at work, and I pay her. I'd be very worried if I didn't have her to do this. I would only take my daughter to family or to a friend I have on another shift who would help; I would not take her to child care."

- **Other arrangements**
  "When I was still married, I changed shifts so my husband (on another fire department) and I could share child care. Unfortunately, there was a lot of stress because he didn't change shifts until 7:00 and I had to be at the station at 7:30. After a while, the person I relieved got tired of standing by for me and wouldn't do it any more. After my divorce, my ex would still come down and get the kids for my 24 hours, but I had to leave them alone in the mornings for 30-40 minutes until he could get there."

  "I drop my daughter off at her family day care on my way to work. She takes the bus to and from school, and goes back to their house. She stays there overnight, and I pick her up in the morning. On weekends, I pretty much depend on close friends to take care of her. Sometimes she stays with her former pre-school teacher, whom I pay for this."

  "My children care for themselves at home; one has taken a babysitting course and is certified in CPR. I work six miles from home... there has been only one home emergency (electrical), and I was relieved immediately by my supervisor. My children know that if they need help, I am only a phone call away. I strongly recommend pagers for custodial parents!"

- **A note from Great Britain**
  "When I joined the fire service, I was the only woman single parent, and initially I was able to claim financial assistance from the fire brigade for child care. This was a small subsidy paid directly into my wages which was later withdrawn after the demise of the (left wing) Greater London Council. It was reintroduced a couple of years later... Although this payment was originally taken up by women, the majority of claimants now are men. The facility is successful if only for the fact that child care is seen as a priority for everyone, not just by women."

Problems and stresses

"I worry about my son being alone so much. After school, he comes home to wait for the babysitter to pick him up; sometimes she works late and doesn't get him until 6:00 or 6:30. I feel guilty for leaving him alone, and also because I make him stay in the house to wait, and he doesn't get to play outside with his friends."

"The first year of a child's growth is very special. Every day something changes, and you're not there to see it or hear it. It's also difficult to take care of normal household chores. Cleaning, cooking, shopping used to be tasks that took seconds; now they take hours."

"My children have suffered greatly. My oldest began acting out due to lack of attention and ran away a lot... Everyone blamed me and my job for the problem... Single moms must be careful not to allow the job to overcome us when we're home. Children can't relate to the stresses of the job, but they do sense our feelings."

"My biggest problem is stress, making arrangements for my daughter to be with someone while I work, especially on overtime when they call at the last minute and I have to find child care. I also feel bad because our changing work days mean I can't put her into any extracurricular programs like gymnastics or soccer, because I'm not there to drive her places all the time. It gets very depressing."

"Child care is by far my biggest problem. My other problem is coping with intense feelings of isolation. I am not sure how much of that is related to my job, but I think it may be inherent in single parenting, and I would guess that my 'abnormal' work schedule also contributes to this."

On the positive side

"Apart from night work, which was the real child care problem, the schedule is an advantage for single parents. We have much more off-duty time to spend with our children than ordinary 9-to-5 workers."

"My daughter came home from the hospital sleeping through the night, so I can spend the evenings doing the bottles and laundry. Also, I have a new partner who is willing to take on some co-parenting roles, and that has helped a lot."

"I think I'm pretty fortunate... Even though the money isn't quite enough, I wouldn't trade the shift work for anything. It allows me so much time with my kids, and time to get all the things done around the house that need attention."

"One of the most positive things about being a single parent was the determination it gave me to succeed. The ability to give birth and bring up a child were of more significance than joining the fire service, and always provided me with the impetus to keep going: 'If I can survive childbirth, I can damn well survive this,' being my motto."

The future

"When my children are both in school full time, I hope to get a second job or further my education. I'll have to see what's happening when the time comes."

"I've been a single mom for 14 years. Now I'm also a grandmother wondering about tomorrow. I'm looking to change my life... I'm good at my job, and it used to make me feel good, too. But it doesn't any more, and I can't seem to get those feelings back."

"I don't work at another job because I've decided that while my daughter is young, I want to be with her as much as possible. Once she starts school, I'll be getting another job to help pay our expenses. This is an expensive place to live, and when you add a child and unconventional work hours, it's down right unaffordable!"

"I hope one day to be able to get into a 9-to-5 job, for several reasons. First, I would like to get custody of my other child back--I gave my ex-husband full custody two years ago, for many reasons--and I don't think I can do it working 56 hours a week. I also want to be home every night and weekends and holidays with my kids. I want them to be able to commit to sports or other activities, knowing Mom can take them to practice at night and be there for games on the weekends. I want them to have a normal life with a mom with normal hours for the first time in their lives. As much as I will miss firefighting and the extra money from my second job (which I'll have to give up), I will get to enjoy my kids and be an active part of their lives. For that I would and will give it all up."

"Over the years my son and I have had our ups and downs, but he now seems to be a reasonably well-adjusted teenager doing well at school, and I've just been promoted for the second time. We must have done something right along the way."

If one single message emerges from these diverse voices, it is that fire departments must address the issue of child care. Failure to do so will mean that parents of both sexes, and particularly single parents, will continue to suffer from unnecessary stress and diminished effectiveness at work, sometimes resulting in their leaving the job. Losses like this serve no one's best interest: not the employee's, not the department's, and certainly not the children's.[6]

Notes:

[1] Women in the Fire Service, Inc., unpublished survey data, 1995.

[2] Only two percent of all government and private employers sponsor day care centers for their employees' children, according to the U.S. Department of Labor's Bureau of Statistics (1988).

[3] Bowers, Karen. "Old firehouses may get rebirth," *Rocky Mountain News*, August 18, 1989; quoting Denver Fire Chief Richard Gonzalez.

[4] Allcock, Ann. "Workplace Child Care and the London Fire Brigade," *WFS Quarterly*, Summer 1991; pp. 8-9.

[5] Stevens, Larry H. "But who will watch the kids?" *Fire Chief*, April 1992, pp. 113-114.

[6] Floren, Terese M. "Firefighter/Single Parent." *Firework*, March, 1994, p. 1.

Hair and grooming standards for firefighters

Hair-length and grooming regulations for fire service personnel have been problematic ever since the first woman firefighter came up against standards that were created originally for men but ended up etched in the stone of fire department tradition. One of the problems in getting fire departments to become more flexible on this issue was the fact that at the time women were beginning to enter the career fire service-- the late 1970's--long hair was fashionable on men. Male firefighters had been agitating for some time to be allowed to wear their hair longer, usually to little avail. When women came forward with the same request, the answer was often a pre-scripted "No." As the fire service in the 1990's finally began to break away from its traditional, authoritarian approach to employees, fire service managers have been adopting more flexible employee grooming standards.

The legal background

Unfortunately for those who would like legal support for their attempts to change fire department hair policies, the courts generally have allowed employers to establish work rules for hair length.

Constitutional protections afforded to employees in this area clearly are limited. Numerous cases have been brought on constitutional grounds, with employees arguing that appearance is an aspect of personal liberty, so that any interference with that "right" must be justified by a legitimate State interest. The courts, in response, generally have upheld a public employer's right to impose grooming regulations that can be justified by the need for discipline, uniformity, and *esprit de corps*, even where no safety considerations exist.[1]

A 1992 case filed under Louisiana constitutional law, however, resulted in a ruling against a public employer. At issue was the city's attempt to restrict firefighters' hair to shoulder-length or shorter. (Women had been on the fire department for 11 years at the time, and previously had been permitted to pin their hair up to conform with the hair standard.) In issuing an injunction against the city, the judge found that "the regulation (was) not gender neutral" because it affected women differently from men and "ha(d) the effect of classification of individuals on the basis of sex." The ruling continued

Similarity of appearance can be and was in fact achieved by requiring fire personnel on duty or in uniform to have their hair, if longer than regulation length, pinned to meet a length requirement established by their employer. The esprit de corps and equal treatment as to grooming standards are easily achieved by uniform enforcement of hair length *while on duty or in uniform* with a recognition of an individual's right to have whatever length of hair he or she desires *as long as while on duty or in uniform it is kept to a level set forth in the employer's regulation*...While similarity of appearance has been recognized as an appropriate and rational goal in a "paramilitary" civilian service, commonality may not and should not be required at the expense of reason and purpose.[2]

Because Louisiana constitutional law differs significantly in this area from that of other States, the ruling is not likely to have a direct impact outside Louisiana.

Many early hair-length cases brought by men, including police officers and firefighters, claimed it was illegal sex discrimination for an employer to have regulations that prohibited long hair on men but allowed it on women. Some fire departments even now continue to insist that they cannot legally have two separate standards for men's and women's hair. Since the mid-1970's, however, the courts have held consistently that prohibiting long hair for male employees only is **not** sex discrimination as long as the employer's grooming standards for both sexes are related to community standards and are applied in an even-handed manner.[3] The courts have made a distinction between hair standards and issues where discrimination affects "fundamental rights" or is based on immutable (unchangeable) characteristics.[4]

The judicial and administrative branches of the government disagree on this basic point of sex discrimination law. The EEOC has consistently treated different appearance rules for men and women as constituting sex discrimination under Title VII.[5] The courts, however, have decided that antidiscrimination law was not intended to interfere with employers who wish to set reasonable appearance standards required by their business.

Safety versus grooming

Apart from the legal aspects, hair-length issues are also confusing because fire departments traditionally have required firefighters to have short hair for two completely different reasons: because, given the protective gear then in use, it was safer during firefighting operations, and because it looked neater, more professional, or more uniform. Where these two reasons have motivated a single policy, it is necessary to sort them out in order to create a new policy that addresses current needs.

A safety standard is gender-neutral: fire will burn exposed hair regardless of the person's sex. Advances in protective equipment in recent years have made it possible for firefighters to have much longer hair than in the past and still be much safer than ever. Self-contained breathing apparatus (SCBA) facepieces, flame-resistant hoods, and flame-resistant helmet liners insure that all surfaces on the head are protected. For this reason, many fire departments have adopted a safety-based hair standard that simply states, "No hair shall be exposed during firefighting activities." This approach avoids the need to regulate the actual length of the hair, and bases its restriction simply on the justifiable need for firefighter safety.

While a safety standard must be gender-neutral, a grooming standard need not be. Even though the EEOC disagrees, the courts have held consistently that employers legally can require male employees to wear their hair shorter than female employees and that such policies do not constitute illegal sex discrimination. Most fire departments that employ women firefighters allow women, if not all employees, to have long hair. Most policies specify that hair longer than a certain length must be restrained or pinned up during all duty hours, or during emergency operations. Policies of this sort usually are achieved through formal or informal negotiations with fire department management, not through legal action.

Notes:

[1] *Quinn v. Muscare*, 425 U.S. 560, *reh'g denied*, 426 U.S. 954 (1976) [male firefighter challenge to hair standard]; *Kelley v. Johnson*, 425 U.S. 238 (1976) [male police officer].

[2] *Sellers v. City of Shreveport, et al.*, 1992. The employer defended the policy as a grooming standard, stipulating that the safety of firefighters was not at issue.

[3] *Dodge v. Giant Food, Inc.*, 488 F.2d 1333 (DC Cir. 1973); *Baker v. Calif. Land Title Co.*, 507 F.2d 895 (9th Cir. 1974) cert de nied, 422 U.S. 1046 (1975); *Longo v. Carlisle DeCoppet & Co.*, 537 F.2d 685 (2d Cir. 1976); *Earwood v. Continental South eastern Lines, Inc.*, 539 F.2d 1349 (4th Cir. 1976); *Barker v. Taft Broadcasting Co.*, 549 F.2d 400 (6th Cir. 1977); *Knott v. Missouri Pac. R.R. Co.*, 527 F.2d 1249 (8th Cir. 1975).

[4] E.g., an employer's refusal to hire women with small children [*Phillips v. Martin-Marietta Corp.*, 400 U.S. 542 (1971)] or the firing of women who marry [*Sprogis v. United Airlines*, 444 F.2d 1194 (7th Cir. 1971)].

[5] E.g., EEOC Dec. No. 71-1529, 3 F.E.P. 952 (5/9/71).

Sample language from fire department hair-length and grooming policies

Following are excerpts from fire department policies regulating employee hair length and the wearing of jewelry. They cover a range of options available to fire service employers for dealing with these questions. Policies should be implemented only with full consideration of your own department's specific needs and after consulting qualified legal advice.

"Hair shall be neatly groomed and the length or bulk of the hair shall not be excessive or present a ragged, unkempt or extreme appearance. (Men:) Hair may cover one half of the ear but shall not cover the entire ear. (Women:) Hair may not extend beyond the lower part of the shoulder blades."

"Members are discouraged from wearing rings or other jewelry on the fire or training ground. Female members may wear earrings providing they do not extend below the bottom of the ear."

(Women:) "Hair must be clean and neatly arranged. When in uniform, back hair must not fall more than one-quarter inch below the lower edge of the collar. No hair must show under the front brim of fire service headgear. Afro, natural, bouffant, and other similar hair styles are permitted, but...bulk of hair must not exceed two inches. In no case is the bulk of the hair permitted interference with the proper wearing of fire service headgear."

"Only pins, combs or barrettes that are similar in color to the individual's hair color may be worn to meet the requirements of the regulation. Jewelry which extends beyond the ear lobe or...is loose or protrudes and may catch in machinery or equipment may not be worn while on duty."

"It is recognized that traditionally acceptable standards for female hairstyles differ considerably from those of males. Female hairstyles that would normally not conform to the standards outlined in the S.O.P. may be pinned up or secured in order to comply while on duty. In these instances, the hair must be pinned up or secured at all times while on duty, and shall not interfere with the proper wearing of uniform hats or protective equipment, or in any way create a safety hazard."

(Men and women:) "There are many hair styles that are acceptable. So long as the person's hair is kept in a neat, clean manner, the acceptability of the style will be judged by these criteria: Hair styles that preclude the proper wearing (of SCBA) are not permitted...Hair will be worn so that is does not extend below the bottom of the uniform shirt color while standing in an erect position. Hair may be pinned or worn in a way to keep hair above the bottom of the collar..."

Sample language from fire department hair-length and grooming policies (cont.)

"To facilitate a professional appearance, hair and grooming standards must be followed. These standards have been modified to meet contemporary styles without jeopardizing the safety of firefighters involved in the hazardous activities associated with firefighting."

"When in a normal standing position, the hair can extend to the top of the collar area. Hair will not extend beyond the bottom of the earlobe. Longer hair is acceptable if it is pinned up in a neat manner and does not interfere with the wearing of departmental headgear. No ribbons or ornaments shall be worn in the hair except for neat inconspicuous bobby pins or conservative barrettes which blend with the hair color. Hair…will not exceed two inches in height."

Fire station facilities

Most fire stations in use today were planned and built with a single-sex workforce in mind. Many of these buildings are now being used by a workforce that includes both women and men. Not surprisingly, this can result in inadequacies that are a source of inconvenience, discomfort, embarrassment, and friction for all concerned.

Fire departments have developed a variety of solutions to problems created by inadequate facilities. The cheapest and easiest answers are usually the first to be implemented: a "Men/Women "or "Occupied/ Unoccupied" flip sign can be installed readily on the door of the station's only restroom or shower, as can a lock on the door. Makeshift partitions, such as a row of lockers or a rollaway curtain, can be put up between beds if bunkroom separation is desired.

These are short-term solutions to real or perceived needs concerning personal privacy in the fire station. The underlying question that guides the development of long-range answers is whether or not men and women on the job should be provided with separate facilities for various personal functions.

The legal background

The expense to the employer of providing separate restrooms, showers, locker rooms, and bunkrooms for women and men usually cannot be used as a reason to exclude women from an occupation.[1] According to the EEOC's guidelines,

> Some states require that separate restrooms be provided for employees of each sex. An employer will be deemed to have engaged in unlawful employment practice if it refuses to hire or otherwise adversely affects the employment opportunities of applicants or employees in order to avoid the provision of such restrooms for persons of that sex.[2]

The Guidelines of the Office of Federal Contract Compliance contain an even more emphatic provision:

> The employer's policies and practices must assure appropriate physical facilities to both sexes. The employer may not refuse to hire men or women, or deny men or women a particular job because there are no restroom or associated facilities, unless the employer is able to show that the construction of the facilities would be unreasonable for such reasons as excessive expense or lack of space.[3]

Since the existence of male-only facilities is often the result of past discrimination, allowing the cost of new facilities as a defense would only honor and perpetuate that discrimination.

One State law that applies to some fire stations is Section 2350 of the California Labor Code. It requires that business establishments that have five or more employees must provide separate bathrooms for each sex, and that no person may use bathrooms designated for the opposite sex. Other States may have comparable provisions in their labor codes or other laws.

Local building or health codes usually require employers to provide bathrooms, and sometimes other facilities, for each gender in the workplace. As most fire stations are the property of municipal or county government, they generally have been made exempt from the provisions of these codes. Where such exemptions do not exist, fire departments would have to comply with the codes.

The impact of inadequate facilities

One observer, commenting on the problem of inadequate fire station facilities, wrote

> Under the best circumstances, bad facilities are an inconvenience which women suffer from in far greater proportion. Under the worst conditions, poor facilities can lead to problems with morale and job performance, and an increase in the occurrence of harassment. At least one discrimination lawsuit has been filed which was due in part to inadequate facilities. A lawsuit costs a lot more than a locker room, and in the end, no one wins.
>
> When the need for women's facilities in the fire station is neither recognized nor addressed, the... department may be saying that women are not important enough here to deserve decent facilities, that women may not be around long enough to warrant planning for the future, that women are not wanted at this station, and this is a reasonable way to keep them out; or that we are too busy here to consider the real needs our personnel. All of these are harmful messages, both for women and for the organization of which they are a part.[4]

Two things usually happen when firefighters in a newly integrated workforce are forced to occupy inadequate facilities. One is that the women are blamed for "causing" the problem. Even though it is the design of the station that is lacking, the feeling among the men is often that since "there wasn't a problem until she got here," it's the woman firefighter's fault. Solutions such as bumping an officer out of his private room to give it to the woman can generate a similar resentment. Where this type of hostility exists, providing facilities that offer privacy for both sexes becomes only half the solution. It is important for management to make it clear that alterations to the facilities are being done not "for the women" but for better privacy for women and men alike.

The other common reaction to inadequate station facilities is the tendency to adapt and accommodate to them as much as possible. Women and men in the workforce--and particularly women, if they are in the minority or are the newest firefighters--usually will adapt to situations that are less than ideal. Many women firefighters do not routinely shower at work, or they get up an extra hour early in the morning in order to shower before the men. Women have learned to use broom closets as changing rooms; firefighters of both sexes develop the habit of looking for feet under the restroom stall walls. But just because a person or group can develop behaviors to cope with a situation or environment doesn't mean it's right to leave things that way indefinitely. All firefighters deserve some basic privacy, either individually or by gender, for personal functions. Providing this privacy should be a management priority.

Solutions

Fire departments should develop 5- or 10-year plans for remodeling inadequate firehouses. All new stations, and any significant remodeling of existing stations should provide adequate facilities for firefighters of both sexes. Most fire stations will stand for a half century or more: it is important to build them not for yesterday's needs or even today's, but tomorrow's.

Although crews in many fire stations manage to cope well enough with shared facilities, it is better if station design provides privacy for both sexes in the restroom, shower, and changing areas. The issue of separate bunkrooms for women and men is more controversial. As mentioned earlier, reassigning an officer's room to a woman firefighter usually creates hard feelings. Tucking an ad hoc "women's bunkroom" off in one corner of the station (such as a rollaway bed in the weight room) is inconvenient for everyone and a clear message that the woman doesn't really belong. The most common solution is for women and men to share the one existing bunkroom. Many women firefighters prefer this arrangement, because it keeps them a part of the crew and a part of the information-sharing process that begins as soon as a call comes in. On the other hand, some men and women are not at all comfortable sharing a bunkroom in this fashion.

The real, long-term solution to bunkroom design is to provide privacy for everyone. Many fire stations now have cubicles containing a bed, desk, lamp, and three or four lockers (for one person on each shift), with a curtain across the doorway. These provide privacy and a reduction of sound and light from coworkers. (Snoring may be a common source of humor among firefighters, but routinely being deprived of sleep by one or more snoring coworkers is also a significant source of job-related stress.) Cubicle partitions typically do not extend all the way to the ceiling, and the open space at the top provides air circulation and allows everyone to hear emergency tones and other broadcast information.

This is a solution that pleases everyone and doesn't pit the women against the men, the paramedics against the suppression personnel, or the officers against the firefighters. It also avoids controversies over whether the women's bunkroom in the new station should be the same size as the men's bunkroom, whether men are allowed to use the women's bunkroom if no women are assigned to the station on that shift, whether a station that houses two officers should have men and women officer's bunkrooms, and so forth. It is a solution that respects the privacy and individuality of all firefighters without regard for gender, and for that reason usually is supported by all concerned.

Notes:

[1] As a rule, expense will not support gender as a bona fide occupational qualification unless the expense would be clearly unreasonable. See EEOC Case No. YNY 9-047 (5/21/69), C.C.H. Empl. Prac. Guide ¶6010.

[2] 29 C.F.R. §1604.2(b)(5).

[3] 41 C.F.R. §60-20.3(e)(1970).

[4] Willing, Linda, "Bedrooms and Bathrooms: The Hidden Message." *WFS Quarterly,* Winter 1988-89, pp. 1-2.

Promotion

Women firefighters often approach promotions with mixed feelings. Although women are as likely as their male counterparts to be interested in career advancement and new challenges at work, the prospect of competing in a promotional process and then starting over in a new position can seem overwhelming. Some women may be reluctant to jeopardize the hard-won acceptance and security they have achieved as firefighters. Others, who have not yet gained that acceptance, are still struggling day to day and can't imagine taking on even more.

Women face a unique set of circumstances when promoted, especially to officers' ranks. In some cases, they must deal with obstacles their male coworkers do not contend with. In other cases, women bring assets to the job they may not even recognize themselves. And in many important ways, women and men share similar fears, hopes, and dreams as they move into advanced ranks in the fire service.

Women considering promotion should remind themselves that men and women are much alike when it comes to apprehension about the new role. Men may express themselves through humor or bravado, while women might sit quietly and worry, but prospective officers of both genders are likely to come up with many of the same fears and insecurities when they honestly confront the issue. Do I know enough? Will my crew trust me? How can I win the respect of the more senior people on the job? What kind of leader will I be? These are questions all new officers must resolve for themselves as they develop and grow in their new roles.

Along with these concerns, women face some obstacles that most men never have to contend with as officers. One of the most frustrating is the likelihood of being treated with disrespect by the public. To many citizens, women do not "look the part" of a firefighter, and such people will find it even less credible that a woman would be in command of a fire crew or fire scene.

Having your authority challenged because of stereotypes or prejudice can be demoralizing. It is tiresome to have to explain constantly that, yes, you are the officer in charge. The situation is made worse if crew members--deliberately or inadvertently--go along with these assumptions, and try to step in as informal leaders.

Similarly, members of the public may treat a woman officer with condescension, assuming her to be less capable than her male coworkers. Men do not have to deal with comments like, "Oh, you can't be a firefighter; you're too small," or "You don't really go into burning buildings, do you, dear?" Men who show up in uniform or fire gear with the fire truck are assumed to be capable because people expect them to be. Some people have different expectations about women in these roles, and may feel free to voice these feelings.

There is little that can be done about how members of the public feel or express themselves, but women officers do have considerable influence over how their own crews react to such incidents. Teambuilding, open communication about the problem, training, and loyalty that is earned will do much to ease the transition for the woman officer.

Women usually lack access to the informal side of the organization that is open to most men. Women are less likely to be included in hunting or fishing trips, golf outings, or socializing after union meetings. In these settings, rank is put aside in favor of camaraderie, and those present can gain tremendous insight into the subtext of an organization. Men also gain important information about coworkers this way. This is one reason men seem to get over the recruit phase of their careers much more quickly than women do.

The solution to this problem is not for a woman to force her presence into unwelcoming environments. A better approach is to develop other ways to gain the same insights. Formal or informal mentoring relationships are an excellent means of doing this. Becoming involved more closely with a smaller number of coworkers through study groups, special teams, or other interest groups also can be very helpful in developing that informal knowledge base. It is also possible to find genuine common interest with co-workers and use that as a springboard to enhanced work relationships. Professional networking is also extremely beneficial.

Perhaps because of lack of access to the "inside story" about promotional roles, women may have unrealistic ideas about what the job entails, and set impossibly high standards for themselves when thinking about meeting those job requirements. When asked if they plan to go for promotion soon, many women say they're not ready, even though they have as many or more years of experience as men seeking promotion. Women tend to think they must be good officers from the start, that anything less would be failure, despite recognizing that men in the role grow with experience and training. Women tend to be hard on themselves, doubting their basic abilities, focusing only on their weaknesses.

It doesn't help that women so often lack role models to guide them. They see only men in the job, and since they are not men, they seem to have failed to live up to the examples just by that fact alone. It is natural for potential officers to want to emulate experienced officers they admire, but this can be tricky when all those officers are men. A woman may feel she must develop a leadership style that is not natural to her, simply because she has seen it work well for others. It is hard for many women to see themselves as officers when they have never seen any women in that role.

Networking with other women firefighters and fire officers is the best solution to this problem. Through Women in the Fire Service, Inc., a national nonprofit organization, and other local and regional groups, women can make face-to-face contact with women who already have traveled the path they are considering. Just meeting a woman captain, shift commander, or chief is a tremendous benefit for a woman firefighter who is wondering if she should take that first promotional step. Seeing real women in those roles, women with widely varying backgrounds and personal styles, makes promotion seem much more real and accessible.

Women often do not see that, despite facing some obstacles, they may bring genuine gifts to the officer's role as well. The greatest of these gifts is simply the unique set of life experiences, wisdom, and knowledge that every individual woman has. Women may have been discouraged from valuing these gifts because they seem irrelevant to the job. Women may have been told, as many recruits are, that being book-smart counts for little if they can't break a door down with a single kick. In reality, this isn't true even for firefighters, and it becomes even less valid with every promotional step a firefighter takes.

It is important to see the promotional process for what it really is. A clear-eyed look at the actual requirements of the job can be quite heartening for many women. A woman who once felt insecure as a firefighter because she didn't know how to tear apart a diesel engine will find written and verbal communications skills, the ability to organize and plan, or skills in teaching are important assets to her as an officer. When women realize they have inherent or acquired abilities that are needed in their new role, they are likely to gain confidence in using those skills and enhancing other traits that may need further development.

Women often say they never feel like real insiders in the fire service, that they can never truly be "one of the guys." It is a fact of life that no matter how assimilated a woman is on the job, at least during this transitional generation, she will always have some objective distance from the job as well. Some women lament this fact, not seeing that as new officers, it is a tremendous advantage. One of the hardest things for male firefighters in becoming officers is the difficulty of drawing the line between being one of the boys and being the leader with authority over people who are so close. Many men struggle with finding ways to create that small but important distance that women come by naturally.

Women certainly do not have a monopoly on intuition, but it is a quality that they have historically been encouraged to develop, and it is an extremely valuable one for fire officers. High-level technical and skills training combined with sensible regard for intuition can be a powerful combination, whether the challenge is commanding a fire scene or handling personnel problems in the station.

Ironically, women sometimes overlook one of the greatest allies in the quest to be credible officers, and that is the rank itself. Many women are unfamiliar or uncomfortable with paramilitary, rank-structured systems. This may get them into trouble as firefighters if they fail to follow the chain of command, and it can also hurt them as officers. Some women do not understand what rank means to most men: that a paramilitary hierarchy demands that respect and regard be paid to rank, no matter who holds it. This means men are likely to show deference to a woman officer by virtue of rank alone, regardless of what they may feel about her personally. This is a real boon to the new woman officer that she should not treat lightly. Women should never discount their rank as unimportant, and should not try to make a fire crew operate like a committee. An officer is expected to be a leader, and women who become officers must accept that role and challenge.

The first step is usually the hardest. Should she take the promotional test or not? Many women waver or give up at this point, often for understandable reasons. Why risk change when things are finally okay? Women may convince themselves that they are better suited for support roles, especially since they see so few (or no) women with real power in the organization. And what about the test? On many fire departments, taking a promotional test is a monumental undertaking, involving months or even years of preparation. It may seem overwhelming, impossible to even attempt.

In fact, women often do have additional pressures on them as they prepare for promotion. Studies show that even in married households where a husband and wife both work full time, the woman still is likely to do most of the home chores and child care. (And single women firefighters are much more likely than their single male counterparts to have children.) Women who work this kind of double duty may be unwilling to take on the demands of preparing for a promotional test.

Men who successfully promote often emphasize how important family support is during this stressful time; this is even more true for women. Women in relationships need to know they have the support of their partners over the long haul. Children need to get involved too, and will be much easier to manage if they feel they are contributing to their mother's success. Women need to evaluate all the demands made on them clearly, and find ways to delegate and modify certain tasks to accommodate the time that they need for study and preparation.

Balance is important, too. Ignoring family and friends for a year in order to become a recluse well-versed in fire facts is not a prescription for a happy life. Good health, good study habits, and a good attitude are all more important than the sheer number of hours put in with the books. It is possible to make a strong commitment to the promotional test without losing one's sanity.

Sometimes women try to find a middle ground by saying they are only taking a promotional test for practice. By saying this, they may feel they are hedging their bets: if they don't study adequately, or fail for some other reason, having said it was for practice gives a good excuse for the negative outcome, making it seem like it really doesn't matter. But approaching a promotional test as if for practice is a bad idea that can lead to all kinds of problems.

The biggest dilemma is, What happens if a practice run results in getting the promotion? In nearly every case, the woman takes the position if it's offered, despite her earlier protests that she was not really ready. This causes understandable resentment among coworkers who were 100-percent committed to the test, and who feel they were defeated by someone who is undeserving. It can also lead to a crisis of confidence for the individual woman who has been telling herself she is not ready for the job, only to find herself suddenly in it.

Women should go for promotions when they are ready to step into the position and do it like everyone else on the job. Women should not feel pressured to take promotional tests because the personnel director or EEO officer or fire chief would like to have a woman in the position. Each woman must decide for herself when she is ready to make the commitment to advanced rank on her department.

But just as women should not be hurried into promotion, they should not wait too long, either. There are always reasons why going for promotion is not the right thing to do. It will never be easy. But women must understand that no one is a good officer from the start, that getting promoted means you have the opportunity to be a beginner again, and that this can be both challenging and hugely satisfying in the long run.

Women should prepare themselves for promotions, and this process can start their first day on the job. The key component to this preparation is training. Women should seek out and take advantage of all the training they can get, both within their department and from outside sources. This can be done at a very informal level, such as staying up to help the captain with his or her report after a call. It might mean volunteering for a special team to gain access to increased training opportunities. It also could include making a commitment to seek out training at seminars and conferences in other geographical areas.

Fire service women's conferences and training seminars at the national level address the needs of those at all levels, with classes specifically aimed at women who are new or aspiring fire officers. Women in the Fire Service sponsors such conferences on a regular basis at various locations throughout the U.S. Other agencies and jurisdictions host similar events at the local and regional level.

Most women find their training is greatly enhanced when it takes place in a supportive and safe environment. For this reason, many women benefit from leadership development training geared specifically toward women. When women do not feel silenced through isolation, different communications styles, or lack of support, they are more free to express their concerns and have their questions answered.

Women should not give up on training even if their departments are not particularly supportive or they don't have the resources to travel to outside seminars. Women gain much from taking a leadership role in developing training opportunities for themselves and others. A woman might organize a mock oral board among test candidates or offer to teach a class in an area of her expertise. She might find ways to network with local firefighters from other departments to share training resources. If she feels she is not in a position to take this kind of initiative on her own, she can form alliances with other interested firefighters on her department, and develop these resources as part of a group.

Access to some form of mentoring relationship greatly helps women considering promotion. A mentor can be anyone on the job who has insight and knowledge about the job and the desire and power to support a woman in her quest for advancement. Women should be wary of trying to involve personal friends in these types of relationships, and should base the relationships instead on professional regard and common interest. For example, a woman interested in rescue specialties might find a mentoring relationship with an experienced rescue officer who admires her desire for increased knowledge and experience. Mentoring relationships are not necessarily easy to form or maintain, but they provide many benefits for both parties.

Women also should realize that they can be mentors to younger firefighters, especially young women coming along on the job. Instead of seeing these women as a threat or indulging in jealousy about how easy these women have it compared to what they themselves went through, women can form a more positive basis for understanding by offering to serve as mentors. Since teaching is often the highest form of learning, women who choose to help others in this way may find they are the ones who gain the most from the relationship.

Women are promoting to fire officers' ranks in increasing numbers each year. Hundreds of women hold the rank of engineer, lieutenant, or captain, and dozens have reached chief-level positions: battalion chief, division chief, assistant chief, and chief of department. Women who promote usually find they have increased opportunity to make a contribution, and thus feel increased job satisfaction. This leads to greater commitment to the organization, manifesting in high job performance, greater respect among coworkers, and increased longevity on the job. A promotion can mean making the move from having a fire service job to having a career. This is a positive step for both the individual and the department.

Women firefighters and aging

Age in our culture is the subject of fear and discomfort, and, therefore, also of humor. "I'm getting too old for this," we groan as we leave our warm beds to make one more fire or ambulance call on a cold night. "Age before beauty," we joke when letting someone pass through a doorway in front of us. "Must be old age," we pretend to excuse ourselves when we can't remember an address, a phone number, or someone's last name. The humor bonds us with others--whatever our differences, we're all in the same boat in this regard--and yet, underneath, the negative attitudes about getting old are undeniable.

This is especially true in the fire service and other jobs that require physical strength, cardiovascular fitness, mental alertness, and quick, reliable recall of information. Our expectations and stereotypes tell us old people are weak, old people have heart problems, old people get senile, old people forget things. When, we wonder, will "old" begin for us? When will this inevitable deterioration take hold enough that we can no longer function well at our jobs?

The doubts and negative expectations are even greater for women. Modern European-American culture has enshrined no positive images of older women beyond those of the sweet little grandmother placidly baking treats for her loved ones. The wise and revered crone, the scarred but strong old woman warrior--these are not part of our mainstream heritage. The woman who starts jogging after retirement and ends up running her first marathon at the age of 74, and the woman who takes up weightlifting at 81 are seen as anomalies, women who are defying the odds, not doing what's expected of them. Old women are expected to be frail, slow-moving, dependent, and housebound.

Age is perhaps the last barrier for women to break in firefighting. Twenty years ago the popular belief was that women couldn't be firefighters. The work of hundreds of women as career and volunteer firefighters gradually disproved that belief. It was then converted into "Women can't have children and go back to being firefighters." Again, the eventual reality shattered the stereotype, as women firefighters had babies and capably returned to active suppression.

Now the myth is that "older" women can't be firefighters. It's a variation that will take a little longer to eliminate through direct evidence. The first wave of women career firefighters came less than 20 years ago; most of these women are only now entering their 40's. Some, though, have reached their 50's; they, along with many women firefighters in the volunteer service, are showing us the reality that gives the lie to the myth.

In 1993, at the age of 74, Frances Furcha was still volunteering as an active firefighter for the Mecklenburg Fire Company in upstate New York. She successfully completed the physically demanding "Initial Fire Attack" State training class when she was 70, as well as many other hands-on firefighting classes. In Boulevard, California, Grace Jepsen became a volunteer firefighter in 1975 when she was 56. Breast cancer and a mastectomy six years later offered only a temporary setback; by the following summer, she was back fighting fires and playing on her women's softball team. She served as chief of the Boulevard Volunteer Fire Department for two years before retiring at the age of 67.

Despite the stereotypes, the reality is that women can be active and competent firefighters, certainly into their 50's and often much longer. The revolutionary idea that age does not determine physical capability was borne out by a 1992 study at the Pennsylvania State University. The study found that "accumulated deficits in abilities are only marginally associated with chronological age," and recommended dropping mandatory firefighter retirements determined by age.[1]

Of course aging has an effect on women, just as it does on men. What is important is to identify and deal with the real effects, rather than buying into and being intimidated by the myths. The Penn State study confirms what common sense dictates: a firefighter who takes care of herself and maintains her fitness is much more likely to be on the job and functioning well at 50 or 60 than one who doesn't. Good nutrition, exercise, healthy life habits, and reduction and management of stress **now**, not later, are the keys. Don't wind up at 60 wishing you'd taken better care of yourself for the past 30 years. Make that investment in your future now.

Some of the deterioration and problems we associate with aging are exactly that: changes our bodies go through as a result of time passing. Others, however, are characterized more accurately as the cumulative effect of years of abuse of our bodies. And even some aspects of the inevitable can be delayed or minimized by taking proper care of ourselves now.

Menopause

The "change of life" marks the end of a woman's natural childbearing years. Different cultures regard this milestone in different ways, many with reverence and respect. Ours has invested it with the stereotype of irrational women driven out of control by hot flashes and hormonal changes and has made it a symbol of old age, not middle life. Not surprisingly, many women react to the prospect of approaching menopause with considerable apprehension.

Menopause, which begins on the average between the ages of 48 and 52, brings two or three changes that may affect women firefighters in their work lives. The only one that all women experience is the cessation of menstrual periods, which in itself can only be an advantage on the job. Most women also experience hot flashes, which may be as mild as a few moments' sensation of warmth, or intense enough to cause profuse sweating and also involve nausea, heart palpitations, and anxiety. They may be more severe during times of stress.

Many women have been able to minimize the impact of hot flashes by keeping active and fit; reducing intake of caffeine, alcohol, sugar, and hot drinks; dressing in layers that can be removed to help regulate body temperature; and by letting those around them know what is going on.

Some women also experience weight changes around the time of menopause that may be related to the body's hormonal changes. Again, exercise and maintenance of fitness can help to counteract any weight gains that are unwanted and unhealthy. Keep in mind that it is just as important for women to keep their body weight up as to keep it down; a weight that is 10 pounds below the ideal weight is riskier, health-wise, than one that is 10 pounds over.

Osteoporosis

The density of human bone reaches its maximum around the time the person is 30 to 35 years old. After that, minerals--particularly calcium and phosphorus--begin leaching from the bones, and the bones become more porous. This decrease of bone density occurs at a rate of about 3 percent every 10 years for both men and women until women reach menopause. After menopause, women's rate of loss increases, on the average, to 3 percent per year.[2] Because women typically have less bone mass to begin with, and up until the present generation have not been aware of ways to prevent or minimize the problem, osteoporosis has its most visible effects on older women.

It is estimated that 15 million women in the U.S. have osteoporosis. Some risk factors are controllable; others are biological or medical. Biological factors include having northern European ancestry, being thin and short, having blonde or red hair and fair skin, being childless, or having had a teenage pregnancy.[3] African-

American women, who have, on average, 10 percent more bone mass than European-American women and possibly additional hormonal protection, are at very low risk for osteoporosis, although the incidence may be increasing.

Controllable risk factors include calcium and vitamin D intake, exercise, smoking, caffeine and alcohol consumption, dieting, and salt intake. A daily intake of 1,000 milligrams (mg) of calcium and 400 I.U. of vitamin D (which allows the body to use the calcium) are recommended for all women. Pregnant women and women over 50 should take 1,500 mg of calcium; nursing mothers 2,000 mg. This is usually best taken in the form of a dietary supplement. "Megadosing" is not recommended: excess calcium cannot be used and will be excreted from the body, while excesses of vitamin D can be toxic.

The purpose of taking calcium is to maximize the density or mass of the bones. Research has shown that bone mass will only increase if demands are placed on it; therefore, it is crucial to do weight-bearing exercise to strengthen the bones. Walking and jogging will increase the mass of the bones only in the lower body. A balanced workout routine--not necessarily a strenuous one--is necessary to develop strength in all areas of the body. Women firefighters, many of whom are very active in a wide range of sports, probably will find they have reduced their risk of osteoporosis considerably. If, however, you have been assigned to a desk job for the past few years or simply have gotten out of the habit of exercising, osteoporosis prevention is an excellent reason to get back into it.

Arteriosclerosis and atherosclerosis

Considered a normal part of aging, arteriosclerosis refers to the gradual loss of elasticity in the arteries of the human body. This is caused by the interaction of certain protein fibers in the blood vessels; the result is a compromising of the effectiveness of the cardiovascular system.

Atherosclerosis, on the other hand, is not a necessary part of aging. It is a build-up of fatty deposits in the arteries that narrows those vessels, again compromising the cardiovascular system and making a complete blockage (heart attack, stroke) more likely. Atherosclerosis is directly linked to dietary intake of saturated fats (mostly animal fats) and to smoking.

Arthritis

Osteoarthritis, or degenerative joint disease, is a localized mechanical disorder considered a usual part of aging: 90 percent of Americans over the age of 50 have it to some degree, though most do not have the joint pain that is its major symptom. Osteoarthritis is caused by the wearing away of the cartilage that protects the ends of the bones at joints such as the elbows, wrists, and knees. Once the cartilage is thinned, the bones can grind against each other, which causes pain and sometimes inflammation. While the causes of this form of arthritis are uncertain, it appears that an inactive lifestyle contributes to it.

Rheumatoid arthritis is a very different disease, an inflammatory autoimmune disorder affecting about 1 to 3 percent of the population. Three-quarters of those who get rheumatoid arthritis are women in their late 20's to early 50's. The disease involves the whole body but affects the joints most, inflaming and damaging them. Early diagnosis is important in order for treatment to be effective.

Sleep demands

Of particular concern to firefighters are the changes in our sleep needs as we get older. Typically, our bodies not only begin to demand more sleep but also become more sensitive to sleep disruptions. This demand runs head-on into the requirements of either career or volunteer firefighting, and it may become harder and harder to wake up clear-headed for that 2 a.m. call or to go back to sleep afterwards. Reducing caffeine intake can help, as can sticking to a regular pattern of sleep on your days and nights off. Most experts warn against napping for those experiencing sleep problems, as it can keep one from being tired enough to sleep at night, but for firefighters and those on similar schedules, a regular routine of napping in the morning after work may be a realistic solution. The nap should be long enough to allow you to feel refreshed, but not so long that it affects your sleep that night. Try to nap in a dark, quiet room (preferably the same place you sleep at night), not on the couch in the living room with the television on.

Other physical effects of aging

As the body gets older, it becomes less efficient at processing substances such as alcohol and other drugs: yet another good reason to reduce or eliminate consumption of these substances. Our bodies also heal less quickly from infection as we age, so we need to pay more attention to small injuries. On the plus side, a benefit for allergy sufferers is that allergic reactions decrease with age.

While no literature exists about older women firefighters, much that can be useful has been written about health, fitness and aging. To make our careers, or our years of volunteer service, as productive as possible, we must all take the time to educate ourselves and then to apply what we've learned to our lives. With good living habits and a little luck, women will be able to combine their years of experience with strong, healthy bodies and be wise and skilled older firefighters, at fifty and beyond.

Notes:
[1] Vance, Robert D., "Alternatives to Chronological Age in Determining Standards of Suitability for Public Safety Jobs," Pennsylvania State University: 1992.

[2] Simkin, Ariel, and Judith Ayalon; *Bone Loading* (London: Priori), 1990.

[3] Doress-Worters, Paula B., and Diana Laskin Siegal, et al., *Ourselves, Growing Older: Women Aging with Knowledge and Power* (New York: Simon & Schuster), 1994.

Resources on women and aging:

The Boston Women's Health Book Collective (BWHBC) publishes an excellent, extensive, and reasonably priced collection of books and literature packets on women's health issues. Their books are available in many women's bookstores; their address is P.O. Box 192, Somerville, MA 02144; phone 617/625-0271; e-mail bwhbc@igc.apc.org

BWHBC's publications include

Ourselves, Growing Older: Women Aging with Knowledge and Power; Doress and Siegal, et al. ($19). This book has a comprehensive "Resources" section listing books, magazines, videotapes, and organizations of interest to women as we age.

The New Our Bodies, Ourselves, BWHBC ($20).

Women's Health (readings on social, economic and political issues); Whatley and Worcester ($17).

"Taking Hormones and Women's Health;" National Women's Health Network ($5).

BWHBC's literature packets range in price from $8 to $20; most are $10 or $15. Topics include: Alcohol, Breast Problems/ Breast Cancer, Fibroids, Hormone Replacement Therapy, Menopause, Mental Health, Premenstrual Syndrome, Smoking, Violence Against Women, and Women of Color Health Issues.

The National Women's Health Resource Center (NWHRC) is a nonprofit organization dedicated to improving women's health. They can provide information on a wide range of health topics by mail or over the phone. Their address is 2440 M St. NW, Suite 325; Washington, DC 20037; phone 202/293-6045. NWHRC's resources include a Women's Wellness videotape series and a collection of information packages.

Breast Cancer (30-minute videotape), $21 to NWHRC members, $31 to nonmembers.

Information packages: Autoimmune Diseases, Breast Cancer ($10/$11), Cardiovascular Disease ($10/$11), Hormone Replacement Therapy, Hysterectomy, Menopause, Osteoporosis, and other topics. $8/$10 each except as noted.

Other women's aging and health resources:

The Black Women's Health Book; Evelyn C. White. Seattle: Seal Press, 1990.

A telephone library operated by the National Menopause Foundation is available at 1-800-MENOASK.

Hot Flash, a quarterly newsletter on women's aging issues, is available from the National Action Forum for Midlife and Older Women, P.O. Box 816, Stony Brook, NY 11790.

Resources

Videotapes and films:

"Intent vs. Impact." Sexual harassment prevention videotape. Available from Anderson Davis, BNA Communications Inc., 9439 Key West Ave., Rockville, MD 20850.

"Meeting the Challenge." 10-minute recruitment videotape for women firefighter candidates. Available from Women in the Fire Service, Inc.; P.O. Box 5446, Madison, WI 53705.

"Sex, Power & the Workplace." 60-minute videotape on sexual harassment, with accompanying resource booklet. Available from Lifeguides/KCET Video; 4401 Sunset Boulevard; Los Angeles, CA 90027; 800/343-4727.

"Trade Secrets: Blue Collar Women Speak Out." 23-minute film on women working in the trades. Available on loan from Chicago Women in Trades, 37 S. Ashland St., Chicago IL 60607; 312/942-1444.

"Valuing Diversity." 5-part videotape series. Available from Copeland and Griggs Productions, 302 23rd Ave., San Francisco, CA 94121. Includes titles such as "Managing Differences," "Diversity at Work," "Communicating Across Cultures," etc.

"What About You?/À toi de choisir!" 19-minute videotape profiling six women working in nontraditional occupations, including a firefighter. Available from Women's Bureau, Labour Canada, Ottawa, Ontario K1A 0J2, Canada; 819/953-0055.

Books and other print resources:

Chetkovich, Carol. *Real Heat: Gender and Race in the Urban Fire Service*. Rutgers University Press, 1997.

Devlin and Associates. *Employment Equity Reference Manual for Ontario Municipal Fire Departments*. Prepared for the Office of the Fire Marshal, Ministry of the Solicitor General, 1991.

FEMA/USFA. *Health and Safety Issues of the Female Emergency Responder*. 1996.

_____. *Stress Management: Model Program for Maintaining Firefighter Well-Being*. 1991.

Loden, Marilyn. *Workforce America! Managing Employee Diversity as a Vital Resource*. Business One Irwin, 1991.

MacKinnon, Catherine A. *Sexual Harassment of Working Women*. Yale University Press, 1979.

Martin, Molly, ed. *Hard-Hatted Women*. Seal Press, 1989.

Petrocelli, William, and B.K. Repo. *Sexual Harassment on the Job*. Nob Press, 1992.

Sanders, Jo Schuchat. *The Nuts and Bolts of NTO: How to Help Women Enter Non-Traditional Occupations*. Scarecrow Press, 1986.

Simons, George, and D. Weissman. *Men and Women: Partners at Work*. Crisp Publications, 1990.

Tannen, Deborah. *You Just Don't Understand: Women and Men in Conversation*. Random House, 1990.

_____. *Talking From 9 to 5*. William Morrow, 1994.

Women's Bureau. *Work and Family Resource Kit*. U.S. Department of Labor, 200 Constitution Ave. NW, Room S-331, Washington, DC 20210. (Single copies available free.)

Women's Issues Advisory Committee. *Guidelines for Integration of Women into the California Fire Service.* California Fire Fighter Joint Apprenticeship Program, 1990.

Organizations and their publications:

International Association of Black Professional Fire Fighters, 8700 Central Avenue; Landover, MD 20785; 301/808-0804. The IABPFF has national, regional and chapter committees on Black Women in the Fire Service.

International Association of Fire Fighters, 1750 New York Ave., NW, Washington, DC 20006; 202/737-8484. The IAFF makes available to its members the IAFF Manual on Human Relations, a "Hair Kit" on fire department grooming standards, the original IAFF/USFA manual, *Managing the Entry of Women in the Fire Service,* and a synopsis of information on reproductive safety, pregnancy, and collective bargaining.

National Association of Hispanic Firefighters, 8035 East R.L. Thornton Freeway, Suite 106; Dallas, TX 75228; phone 214/327-8161; e-mail: nahfnp@aol.com. The NAHF has chapters in many States.

9 to 5, National Association of Working Women, 614 Superior Ave., NW; Cleveland, OH 44113; 216/566-9308. Job Problem Hotline: 800/522-0925 (from Ohio: 216/621-9449). Resources and guidance for women on sexual harassment and other work-related concerns.

NOW Legal Defense and Education Fund, 99 Hudson St., New York, NY 10013; 212/925-6635. Information on sexual harassment; guidance on antiharassment policy development.

Women in the Fire Service, Inc., P.O. Box 5446, Madison, WI 53705; 608/233-4768; fax 608/233-4879; e-mail: info@wfsi.org. WFS publishes a quarterly periodical on fire service women's issues, women firefighters' recruitment literature, and information packets on a wide range of issues. WFS also holds conferences on fire service women's issues and a biennial leadership training seminar.

Women's Legal Defense Fund, 1875 Connecticut Ave. NW, Suite 710; Washington, DC 20009; 202/986-2600. Information on sexual harassment; advocacy on harassment and other sex discrimination issues.

Bibliography

Affirmative action

"DCFD Affirmative Action Hiring Plan Overturned." *Fire Chief*, May 1987: 28.

"Female Firefighters Decline Affirmative Action Promotions." *Fire Engineering*, 145(5) (May 1992): 16.

Osby, Robert E. "Guidelines for Effective Fire Service Affirmative Action." *Fire Chief*, Sept. 1991: 50-54.

Schumacher, Joe. "Affirmative Action Revisited." *Fire Chief*, Mar. 1989: 51-53.

Slack, James D. "Women, Minorities, and Public Employer Attitudes: The Case of Fire Chiefs and Affirmative Action." *Public Administration Quarterly*, 13(2) Fall 1989: 388-411.

Family issues

"London Fire Brigade Considers Child Care Alternatives." *IAFC On Scene*, Sept. 15, 1991: 1-2. [Condensed from an article in *Women in the Fire Service Quarterly*, Summer 1991.]

Willing, Linda. "Love on the Job." *Fire Chief*, Aug. 1990: 92.

Firefighter training and pre-training programs

Bird, James W. "Training Women for the P.A.T." *Fire Engineering*, Mar. 1991: 87-93.

"Eleven Women Survive NYFD Training Program." *International Fire Chief*, 49(8) Jan. 1983: 8.

Larkin, Susan R. "Training for Success." *Fire Command*, Aug. 1989: 38-42. [Training women for FDNY entry-level testing.]

McDonald, Bernie R. "Pre-Employment Training: One Department's Program." *Fire Command*, Aug. 1987: 24-28. [Louisville Fire Department]

Pletan, Owen D. "A Format for Successful Pre-recruit Training: The Seattle Method." *Fire Command*, 48(8) Aug. 1981: 35-37.

"Women's Training Program in Jacksonville." *Fire Chief*, Feb. 1991: 60-61.

"Women's Training Program Upgrades Firefighting Skills." *Fire Chief*, 35(2) Feb. 1991: 60-61.

Legal issues

"Department of Justice Consent Decree." *Fire Chief*, July 1987: 30.

"Newest Court Ruling Backs Female Hires." *International Fire Fighter*, 70(5-6) May-June 1987: 1.

Rukavina, John. "Fire Service, Meet the ADA." *Fire Chief*, June 1992: 30+. [Civil Rights Act of 1991]

_____. "Seeing the Future." *Fire Chief*, Sept. 1992: 22+. [Regarding Pennsylvania State University report on the pending expiration of the Age Discrimination in Employment Act exemption for firefighters.]

Schmidt, Wayne W. "Background Investigations." *Fire Chief*, July 1990: 18. [Limits on background investigations in areas such as sexual activity.]

_____. "Civil Liability for Wrongful Discharge," *Fire Chief*, Sept. 1989: 26+.

_____. "Nepotism and Consanguinity Relations." *Fire Chief*, Feb. 1985: 14. [9th Circuit affirms lower court ruling that one employee may be required to transfer or quit if two employees marry.]

Shearer, Robert W. "Can After-Hours Conduct Be Grounds for Firing?" *Fire Chief*, May 1989: 59-60.

"US Court of Appeals Throws Out FDNY Scoring System." *Fire Chief*, May 1987: 14-15.

Physical fitness and physical performance testing

Bell, Laura. "Where Does Physical Testing Leave Women?" *Management Review*, Dec. 1987: 47-50.

Clark, Allen. "Female Firefighters Not the Issue--It's Physical Fitness." *Fire Chief*, 35(10) Oct. 1991: 22.

Davis, James E. "A Look at Performance Standards." *Fire Chief*, Aug. 1991: 56.

Doolittle, T.L. "Validation of Physical Requirements for Firefighters." Seattle: Seattle Fire Department, Office of Management and Budget, 1979.

Evans, D.H. "Height, Weight and Physical Agility Requirements." *Journal of Police Science and Administration*, Dec. 1980: 414-436.

FEMA/USFA. *Physical Fitness Coordinator's Manual for Fire Departments.* 1990.

George, Arthur E. "Only One Standard." *Fire Engineering*, 141(3): 37-38.

Kay, Herbert. "Testing Recruits." *Fire Chief*, Apr. 1989: 70+.

Misner, J.E., S.A. Plowman, and R.A. Bioleau. "Performance Differences between Males and Females on Simulated Firefighting Tasks." *Journal of Occupational Medicine*, 29, (1987): 801-805.

Rafilson, Fred M. "Legislative Impact on Fire Service Physical Fitness Testing." *Fire Engineering*, 148(4) Apr. 1995: 83-84.

Schmidt, Wayne W. "Physical Fitness Standards." *Fire Chief*, July 1989: 42. [Legality of employment requirements restricting body fat.]

Williams, Timothy, & S. Evenson. "Physically Fit For Duty? By Whose Standards?" *Fire Chief*, Mar. 1988: 43+; Apr. 1988: 58+; May 1988: 55+.

Promotion

Enbysk, Liz Peeples. "First Female Fire Chief in Paid Position." *American Fire Journal*, 35(10) Oct. 1983: 14.

Glass Ceiling Commission. *A Solid Investment: Making Full Use of the Nation's Human Capital: Recommendations of the Federal Glass Ceiling Commission.* Washington, DC, Nov. 1995.

_____. *Good For Business: Making Full Use of the Nation's Human Capital; The Environmental Scan.* Washington, DC, Mar. 1995.

Hirschman, Jessica. "Climbing the Glass Ladder--Part II. *Firefighter's News*, 11(3) June-July 1993: 44-47.

Protective clothing

Duffy, Richard; J. Sawicki, and A. Beer. "Project FIRES: Final Report." International Association of Fire Fighters, 1985.

National Fire Protection Association. "NFPA 1971: Standard on Protective Clothing for Structural Fire Fighting." 1991.

_____. "NFPA 1972: Standard on Helmets for Structural Fire Fighting." 1987.

_____. "NFPA 1973: Standard on Gloves for Structural Fire Fighting." 1988.

_____. "NFPA 1974: Standard on Protective Footwear for Structural Fire Fighting." 1987.

_____. "NFPA 1975: Standard on Station/Work Uniforms for Structural Fire Fighters. " 1990.

_____. "NFPA 1981: Standard on Open-Circuit Self-Contained Breathing Apparatus for Structural Fire Fighting." 1987.

Neeves, R., *et al.*,"Physiological and Biomechanical Changes in Fire Fighters Due to Boot Design Modifications." International Association of Fire Fighters and the Federal Emergency Management Agency, 1989.

Sylvia, Dick. "New Turnout Gear, Women's Roles Among Topics at IAFC Conference." Fire Engineering, 131(11) Nov. 1978: 58-62.

Recruitment and retention

Bifano, Angie. "Firefighter Selection Today: The Problem of Balancing Legal, Social and Occupational Safety Issues." Firefighter's News, 9(5) Aug.-Sept. 1991: 48-50.

Booth, Walter. "Recruiting Women and Minorities." Fire Chief, May 1987: 49-53. [Survey of large fire departments regarding their recruitment strategies.]

Brown, Marsha D. "Getting and Keeping Women in Nontraditional Careers." Public Personnel Management Journal, Winter 1981: 408-411.

Durkin, Edward D. "Recruiting and Hiring Women Firefighters." Fire Chief, May 1981: 52-55.

Goldfeder, William. "Retaining and Recruiting Members." Fire Engineering, May 1992: 10-13.

Hammond, Ken. "Recruiting Women Firefighters." Fire Chief, Oct. 1987: 40-41.

Makela, William. "Women Taking Up the Slack in Recruiting." Minnesota Fire Chief, 27(1) Sept.-Oct. 1990: 12-13, 59.

Marinucci, Richard A. "Attracting Recruits: A Matter of Image. "Fire Engineering, July 1991: 10. [Recruiting volunteer firefighters.]

Sanders, Jo Schuchat. The Nuts and Bolts of NTO: How to Help Women Enter Non-Traditional Occupations. Scarecrow Press, 1986.

Scotford, Garth. "Program Revealed Problems in Attracting Women Recruits." Fire, 82(1019) May 1990: 9. [Letter to the editor.]

Thaut, Stanley L. "A History of Tacoma's Effort to Recruit Women Firefighters." *Fire Chief*, 23(9) Sept. 1979: 40-43.

Waters, Michael S. "The Recruitment and Retention of Women in the Career Fire Service." *International Fire Chief*, May 1986: 14-17.

_____. "The Recruitment and Retention of Women in the Career Fire Service, Part II." *International Fire Chief*, June 1986: 18-21.

_____. "The Recruitment and Retention of Women in the Career Fire Service, Part III." *International Fire Chief*, July 1986: 20-23.

Reproductive safety

Dixon, Ernest M. "Reproductive Disorders of the Male/Female Worker: Occupational Placement of Women of Reproductive Capacity--Views of the Medical Community." Occupational Safety and Health Symposia, 1978: 27-28. Cincinnati, Ohio: U.S. National Institute for Occupational Safety and Health, June 1979.

Evanoff, Bradley A. and Linda Rosenstock. "Reproductive Hazards in the Workplace: A Case Study of Women Firefighters." *American Journal of Industrial Medicine*, 9(6) 1986: 503-515.

Fischer, David R., William A. Jones, Charles A. Lacroix, Clay A. Phillips, Perry E. Ray, and Timothy P. Travers. *Written Policies and Pregnant Firefighters*. Emmitsburg, MD: National Fire Academy, June 10-21, 1991.

Infante, Peter F. "Reproductive Disorders of the Male/Female Worker: Occupational Placement of Women of Reproductive Capacity--OSHA's View." Occupational Safety and Health Symposia, 1978: 29-30. Cincinnati: U.S. National Institute for Occupational Safety and Health, June 1979.

McDiarmid, Melissa, M.D., et al. "Reproductive Hazards of Firefighting I and II." *American Journal of Industrial Medicine*, 1991: 433-472.

Olshan, Andrew F., K. Teschke, and P. Baird. "Birth Defects Among Offspring of Firemen." *American Journal of Epidemiology*, 131(2): 312-321.

Stellman, Jeanne M. "Reproductive Disorders of the Male/Female Worker: The Effects of Toxic Agents on Reproduction." Occupational Safety and Health Symposia, 1978: 16-26, Cincinnati, Ohio: U.S. National Institute for Occupational Safety and Health, June 1979.

Templeton, Randy. "Pregnant Firefighter." *Fire Chief*, 36(4) Apr. 1992: 116-118, 120.

Sex discrimination/sexual harassment

Barron, Donna. "Men, Women and Harassment." *Emergency*, 20(7) July 1988: 31-35.

Blackistone, Steve. "Sex Discrimination." *Firehouse*, 15(11) Nov. 1990: 69.

_____. "Sexual Harassment in the Firehouse (Fire Law column)." *Firehouse*, 17(4) Apr. 1992: 90-92.

Beekman, Peter and Allen Fankhauser. "Like It or Not, Sex Is Here to Stay." *JEMS*, 16(1) Jan. 1991: 13-14.

"Cleveland FD Charged With Discrimination." *International Fire Chief*, 50(2) Feb. 1984: 10.

Farley, Lin. *Sexual Shakedown: The Sexual Harassment of Women on the Job*. McGraw-Hill, 1978.

"Firefighters Transferred in Queens Sex-Bias Case." *Fire Control Digest*, 17(11) Nov. 1991: 6-7.

"Justice Actions." *Fire Chief*, Nov. 1986: 25. [Charleston, WV, Police Department ordered to reinstate dispatcher fired when she became pregnant.]

"Justice Department Acts in Discrimination Cases." *Fire Chief*, July 1985: 8-10. [Deletes numerical hiring goals implemented in 1977 consent decree in San Diego.]

"Justice Department Sues City." *Fire Chief*, Nov. 1991: 38. [Dept. of Justice alleges discrimination against white men when West Palm Beach directed the hiring of minorities and women to fill twelve vacancies.]

MacKinnon, Catherine A. *Sexual Harassment of Working Women*. Yale University Press, 1979.

McQueen, Iris. "Sexual Harassment." *Fire Chief*, Aug. 1985: 69-72.

Moore, Thomas V. "Sexual Harassment in the Firehouse?" *Firehouse*, 10(18) Aug. 1985: 14.

Naczi, Frances D. "Removing Sexism From Communications." *Fire Chief*, Nov. 1984: 45-46.

Paludi, Michele A., and R.B. Barickman. *Academic and Workplace Sexual Harassment*. State Univ. of New York, 1991.

Petrocelli, William, and B.K. Repa. *Sexual Harassment on the Job*. Nolo Press, 1992.

Randleman, William. "What is Discrimination?" *Fire Chief*, Nov. 1984: 27.

Schmidt, Wayne W. "Sex Discrimination." *Fire Chief*, 26(4) Apr. 1982: 16.

_____. "Quotas and Other Discrimination Remedies." *Fire Chief*, Apr. 1985: 14.

_____. "Sexual Harassment." *Fire Chief*, Oct. 1986: 14-15. [U.S. Supreme Court decision in *Meritor v. Vinson*; also Indiana case where judge found sexual harassment but not sex discrimination.]

_____. "Sexual Harassment." *Fire Chief*, Feb. 1987: 27. [Appeals court reversal of lower court decision that failed to find harassment evidence of discrimination.]

_____. "Sexual Harassment." *Fire Chief*, Dec. 1991: 30. [Cases involving women police officers.]

Schrader, George. "Avoid Sexual Harassment Hassles." *Fire Chief*, June 1990: 47+. [Sexual harassment not an issue of sex but of power and control; recent court cases.]

"Sex Discrimination." *Fire and Police Personnel Reporter*, Aug. 1984: 13-15.

"Sexual Harassment." *Fire and Police Personnel Reporter*, Oct. 1986: 12-14.

Shouldis, William. "Sexual Harassment." *Fire Engineering*, Sept. 1991: 101+.

Siegel, Deborah L. *Sexual Harassment: Research and Resources*. National Council for Research on Women, 1991.

Webster, Cindy. "Facing Off On Sexual Harassment." *Fire Chief*, Aug. 1992: 72-77.

Supporting workforce diversity

Belenky, Mary Field, et al.;Women'sWays of Knowing, Basic Books, 1986.

Bell, Derrick, Faces at the Bottom of theWell. Basic Books, 1992.

Brightmire, Susan. "Diversity." Fire Engineering, 148(1) Jan. 1995: 117-118.

Briese, Gary. "The Challenge of the '90s: Prepare for the 21st Century." The Minnesota Fire Chief, 28(4) Mar./Apr. 1992: 14-15.

Cannon, Katie G., BlackWomanist Ethics, Scholars Press, 1988.

Cary, Lorene, Black Ice, Knopf (Random House), 1991.

Gilligan, Carol, In a DifferentVoice, Harvard University Press, 1982.

Helgesen, Sally, The Female Advantage, Doubleday, 1990.

Johnston, William B. Workforce 2000: Work andWorkers for the 21st Century. Hudson Institute, 1987.

Katz, Judith H., White Awareness: Handbook for Anti-Racism Training, University of Oklahoma Press, 1978.

Kegan, Robert, In Over Our Heads, Harvard University Press, 1994.

Kochman, Thomas, Black & White: Styles in Conflict, University of Chicago Press, 1981.

Lightfoot, Sara Lawrence, Balm in Gilead, Addison-Wesley Pub. Co., 1988.

_____. I've Known Rivers, Addison-Wesley Pub. Co., 1994.

Loden, Marilyn, Workforce America! Managing Employee Diversity as aVital Resource, Business One Irwin, 1991.

Macklin, Victoria S. "Peer Mediation Helps Heal a House Divided." NFPA Journal, May-June 1991: 62-66.

Morrison, Ann M., Breaking the Glass Ceiling, Addison-Wesley Pub. Co., 1992.

_____. The New Leaders: Guidelines on Leadership Diversity in America, Jossey-Bass, 1992.

Perry, Linda A.M., Turner, Lynn H. & Sterk, Helen M. (eds.); Constructing and Reconstructing Gender, SUNY Press, 1992.

Simons, George. Working Together: How to Become More Effective in a Multicultural Organization. Crisp Publications, 1989.

Simons, George, and D.Weissman. Men andWomen: Partners atWork. Crisp Publications, 1990.

Smith, Michael H. "Communications Skills for a Changing Fire Service." Fire Chief, Sept. 1991: 82.

Sturzenacker, Gloria. "Prejudice Prevention." Chief Fire Executive, Apr.-May 1986: 43-51.

Takaki, Ronald T., A Different Mirror, Little, Brown & Co., 1993.

Tannen, Deborah, Gender and Discourse. Oxford University Press, 1994.

_____. Talking From 9 to 5. William Morrow, 1994.

_____. You Just Don't Understand: Women and Men in Conversation. Ballantine Books, 1990.

Three Rivers, Umoja; *Cultural Etiquette*, Market Wimmin, 1990.

Volunteer firefighters

Cashman, John R. "Gal Fire Fighters Do What's Needed in Brigade Too Small for Specialists." *Fire Engineering*, 132(12) Dec. 1979: 14-16.

Chambers, Mary D. "Volunteer Fire Chiefing." *International Fire Chief*, Aug. 1980: 15.

Marinucci, Richard A. "Women in the Volunteer Fire Service." *Fire Engineering*, Jan. 1991: 10-12.

Mitchell, Carol Ann. "A History of Service: Women in Volunteer Fire Companies." *California Fire Service*, 2(6) June 1991: 24.

Perkins, Kenneth B. "Volunteer Fire Fighters in the U.S: A Sociological Profile of America's Bravest." National Volunteer Fire Council, 1987.

Women firefighters

"All Female Fire Brigade." *Rekindle*. 14(9) Sept. 1995: 5.

Beaver, Don. R and Jerry Knapp. "Women in the Fire Service." *Fire Command*, 51(8): 15.

Casey, Jim. "Women Fire Fighters Are Here to Stay." *Fire Engineering*, 131(3) Mar. 1978: 6.

Chambers, Mary D. "Women in the Fire Service." *Western Fire Journal*, Apr. 1981: 40.

Chapman, Brenda J. "Women in the Fire Service: Do We Belong?" 20(5) May 1991: 11.

Clayton, Bill. "Female Inmates Work as Wildland Firefighters." *American Fire Journal*, 36(10) Oct. 1984: 45-46.

Cridland, Elizabeth. "Ideas for Increasing Women's Role in Fire Service Voiced at Seminar." *Fire Engineering*, 134(4) Apr. 1980: 28, 30.

DeMars, Denise. "What's the Big Deal? (Thoughts of an Individual Fire Fighter)." *Fire Command*, 51(8) Aug. 1984: 23.

Dernocoeur, Kate, EMT-P, and James N. Eastman, Jr., ScD. "Have We Really Come a Long Way? Women in EMS Survey Results." *JEMS*, 17(2) Feb. 1992: 18-19.

Dessoff, Alan L. "Female Inmates: No-Holds-Barred Brigade." *Firehouse*, 48(8) Aug. 1981: 35-37.

Feldman, Danah. "Wildland Fire Fighting." *International Fire Chief*, 46(8) Aug. 1980: 16-17.

"Fire Division Gets First Female Apparatus Operator." United Press International, Regional News-- Cincinnati, OH. (January 6, 1992).

Floren, Terese M. "1990 Survey Results." *WFS Quarterly*, Winter 1990-1991: 14-17.

_____. "Women Firefighters: The Chief's Role." *Fire Chief*, May 1981: 48-51.

_____. "Women Firefighters Speak: A Survey of the Nation's Female Firefighters." *Fire Command*, Dec. 1980: 22-24; Jan. 1981: 22-25.

Granito, Dolores. "More Women Entering Fire Service." *Fire Engineering*, vol. 131, Mar. 1978: 29-30.

Hallinan, Lorin. "Breaking the Barriers." *Emergency*, 26(5) May 1994: 32-37.

Hamilton, Jo Carol. "Women in the Fire Service." *Fire Chief*, 22(8) Aug. 1978: 81-84.

Istvan, Sharon. "Fire Protection Engineering." *International Fire Chief*, Aug. 1980: 21.

Keene, Kathy. "What Is It Like To Be a Female Firefighter?" *Fire Chief*, Sept. 1991: 72-74.

Keller, D. F. "Women Prisoners Protect Facility, Serve Outside Area." *Fire Engineering*, 132(8) Aug. 1979: 124.

Knapp, Jerry and Don R. Beaver. "Women in the Fire Service: Two Views" *Fire Command.*, no. 8 Aug. 1984: 15.

Lipkin, Harriett. "Smoothing the Way for Women." *International Fire Chief*, Aug. 1980: 22-24.

Love, Myron. "Still a Tough Job, But No Longer a 'Man's' World'." *Fire Fighting in Canada.*, 36(2) Mar. 1992: 6-7.

"Niche Jobs Can Be Found for Women." *Fire*, 82(1018) Apr. 1990: 25.

"Oregon Volunteer Firefighter of the Year: Captain Mary Lou Fletcher." *Fire Command*, Sept. 1988: 6.

Orr, Robert. "Women's Work?" *Fire Prevention*, no. 244 Nov. 1991: 10.

Pantoga, Fritzie. "Women Firefighters--A Survey." *Fire Chief*, Jan. 1977: 51-54.

Perkins, Kenneth B. "Women in the Ranks." *Firehouse*, 8(3) Mar. 1983: 49-50.

Quinn, Richard C. "First Woman Fire Fighter." *Fire Command*, 47(9) Sept. 1980: 29.

Roche, Diane C. "Public Fire Education." *International Fire Chief*, 46(8) Aug. 1980: 20-21.

Rudder, Beatrice. "Career Fire Fighting." *International Fire Chief*, 46(8) Aug. 1980: 16.

Rule, Charles H., R. E. Osby, J.H. Steffens, and M.R. Rakestraw. "Workforce 2000." *Fire Chief*, Jan. 1991: 36-40.

Senk, Terry A. "The Firefighter is a Lady--Women in the Fire Service." *The Voice*, 20(4) Apr. 1991: 8-9.

Sexson, Margarita Y. "Fire Departments Surveyed on Employment of Females." *Fire Engineering*, 134(6) June 1981: 48-49.

Smith, Dennis. "Women in the Fire Service." *Firehouse*, 8(2) Feb. 1983: 5

Summers, Liz. "We're Here To Stay: Reflections of a Woman in the Fire Service." *Voice*, 24(5) June 1995: 41-42.

Swartout, Robert. "Women Fire Fighters: The Seattle Concept." *International Fire Chief*, Aug. 1980, 10-11.

Thompson, F. McKeen. "Woman's Place." *Emergency*, 19(4) Apr. 198: 4.

Townley, John P. "Hiring Women Fire Fighters Opposed." *Fire Engineering*, 131(3) Mar. 1978: 33.

Walker, Leslie. "A Firefighter Like You." *Voice*, 24(5) June 1995: 43-44.

Wasylyk, Sylvia. "Women in the Fire Service--Issues and Answers." *Voice*, 23(8) Sept. 1994: 38-41.

Wauls, Bonita. "The Battle's Over." *Fire Command*, 51(10) Oct. 1984: 6. [Letter to the Editor.]

Webster, Cindy. "First National Conference of Fire Service Women." *Fire Chief*, Feb. 1986: 44-47.

Winkle, William, and R. Navarre. "Females in the Fire Service: The Process of Acceptance." *Fire Chief*, Apr. 1985: 68-69. [How the Toledo Fire Department integrated women into its ranks.]

"Woman Loses 9th Try to Join Fire Department." *Fire Control Digest*, 17(1) Jan. 1991: 10.

"Women Are Fire Fighters, Too!" *Fire Command*, Feb. 1976: 16-19.

"Women in the Fire Service." *Fire Chief*, May 1981: 48-51.

"Women in the Fire Service: The Challenge in America." *Fire International*, 7(73) Dec. 1981: 41-43.

"Women Firefighters Still Struggle." *Fire Chief*, 31(3) Mar. 1987: 6.

Women's Issues Advisory Committee. *Guidelines for Integration of Women into the California Fire Service*, California Fire Fighter Joint Apprenticeship Program, 1990.

Ziolkowski, Heidi M "Vying for a 'Man's Job' Now More Than Wishing on a Star." *Western Fire Journal*, 35(8) Aug. 1983: 5.

Other

Buchbinder, Laura B. and Carol Vougioukles. "Fire Administration Initiates Women's Program." *International Fire Chief*, 46(8) Aug. 1980: 12-13.

Burton, Mike. "In a Family Sort of Way." 24(5) June 1995: 17-20.

Chetkovich, Carol. *Real Heat: Race and Gender in the Urban Fire Service*. Rutgers University Press, 1997.

Craig, Jane, and R. Jacobs. "The Effect of Working with Women on Male Attitudes toward Female Firefighters." *Basic and Applied Psychology*, Mar. 1985: 61-74.

Deasy, M. "One Size (Does Not) Fit All." *Firehouse*, May 1988: 33-36.

Devlin & Associates, *Employment Equity Reference Manual for Ontario Municipal Fire Departments*, prepared for the Office of the Fire Marshal, Ministry of the Solicitor General, 1991.

Fire Department Personnel Management Handbook. Managing the Entry of Women and Minorities. Washington, DC: FEMA, 1982.

Guidelines for Integration of Women into the California Fire Service. CA: California Fire Fighter Joint Apprenticeship Committee's Women's Issues Advisory Committee, 1990.

Managing Fire Services. Ronny J. Coleman and John A. Granito, Eds. p. 230-231, 266-269. Second ed. Washington, DC: International City Management Association, © 1988.

Martin, Molly, ed. *Hard-Hatted Women*. Seal Press, 1989.

McCarl, Robert. *The District of Columbia Fire Fighters Project: A Case Study in Occupational Folklife*. Smithsonian Institute Press, 1985.

McNichol, J., and S. Scanlin. "Proceedings of the National Firefighter Health and Safety Forum." Congressional Fire Services Institute, 1991.

National Fire Protection Association. "NFPA 1500: Standard on Fire Department Occupational Safety and Health Programs." 1987.

Navarre, Raymond J. "Developing a Stress-Reducing Fire Station." *Fire Chief*, Feb. 1987: 46-48.

Paradigm, Inc. *Issues for Women in the Fire Service*. Washington. DC: Fire Administration, Sept. 1980.

Stress Management: Model Program for Maintaining Firefighter Well-Being, Washington, DC: FEMA, 1991.

Thomason, Betsy. "Self-discovery: A Way to Deal With Stress." *Fire Chief*, Feb. 1991: 25.

Tokle, Gary. "1001 Considerations." *Fire Command*, Apr. 1989: 24-25.

Turner, Gary. "Butting Heads over Change." *Chief Fire Executive*, Jan.-Feb. 1987: 27-30.

Vonada, Michael. "Shadow Dancing." *Fire Chief*, Dec. 1987: 50.

Photo credits:

www.ingramcontent.com/pod-product-compliance
Lightning Source LLC
Chambersburg PA
CBHW081549170526
45166CB00009B/2631